Rationed Life

Published in 2016 by
Berghahn Books
www.berghahnbooks.com

Library of Congress Cataloging-in-Publication Data

Names: Kučera, Rudolf, 1980–
Title: Rationed life: science, everyday life, and working-class politics in the Bohemian lands, 1914–1918 / Rudolph Kučera.
Description: New York : Berghahn Books, 2016. | "Revised and extended manuscript that was originally published in Czech in 2013"—Acknowledgments. | Includes bibliographical references and index.
Identifiers: LCCN 2015042121 (print) | LCCN 2015048392 (ebook) | ISBN 9781785331282 (hardback : alkaline paper) | ISBN 9781785331299 (ebook)
Subjects: LCSH: World War, 1914-1918--Social aspects—Czech Republic—Bohemia. | Rationing--Social aspects—Czech Republic—Bohemia—History—20th century. | Science—Social aspects—Czech Republic—Bohemia—History—20th century. | Working class—Czech Republic—Bohemia—History—20th century. | Food—Political aspects—Czech Republic—Bohemia—History—20th century. | Labor—Political aspects—Czech Republic—Bohemia—History—20th century. | Sex role—Political aspects—Czech Republic—Bohemia—History—20th century. |Protest movements—Czech Republic—Bohemia—History—20th century. | War and society—Czech Republic—Bohemia--History--20th century. | Bohemia (Czech Republic) —Social conditions—20th century.
Classification: LCC D524.7.C94 K83 2016 (print) | LCC D524.7.C94 (ebook) | DDC 943.71/024—dc23
LC record available at http://lccn.loc.gov/2015042121

British Library Cataloguing in Publication Data

A catalogue record for this book is available from the British Library

ISBN 978-1-78533-128-2 hardback
ISBN 978-1-78533-129-9 ebook

Contents

Acknowledgments vi

Introduction 1

Chapter 1. Rationed Satiety: The Politics of Food 12

Chapter 2. Rationed Fatigue: The Politics of Work 57

Chapter 3. Rationed Manliness: The Politics of Gender 94

Chapter 4. Rationed Anger: The Politics of Protest 130

Conclusion 163

Bibliography 171

Index 187

Acknowledgments

This book is a revised and extended manuscript that was originally published in Czech in 2013. Neither its Czech nor its English versions would have been possible without the kind and selfless help of many friends and colleagues. I would like to extend my sincere thanks to Ota Konrád, Vítězslav Sommer, and Jakub Rákosník, who were the first critical readers of the whole manuscript and significantly contributed to its final shape. Historians Dagmar Hájková and Josef Tomeš were always ready to answer any questions no matter how meaningless, albeit of vital importance to me. I am indebted to Jakub Beneš, whose close readings and comments on both language versions were invaluable. I am also grateful to Nancy Wingfield for her tireless encouragement during the arduous task of revising the English manuscript.

As I wrote each chapter of the book, I tried to stay in close dialogue with the scholarly community. I thank all those who discussed my arguments at various stages of their development at conferences all over the Czech Republic and Europe. Especially the 6th Conference of the International Society for First World War Studies that took place in 2011 in Innsbruck was a great source of inspiration in this context. Presenting my research and discussing it helped clarify several questions, which led to my first ideas on how the book would be structured. I am grateful in particular to Gunda Barth-Scalmani, Wolfram Dornik, and Mark Cornwall for their suggestions and comments.

The research that preceded the writing of the book would not have been possible without the exceptional kindness of many institutions that opened their doors and archives to me. I must name especially the National Library of the Czech Republic, the Military Historical Archives of the Czech Republic, the Labor Union Archives, and the Prague Municipal Archives. Hana Svatošová, Veronika Knotková, and Martina Power deserve special mention here. Without their hospitality and helpfulness, many sources that have proven invaluable for my book would have remained out of my reach. My research was generously supported by the Masaryk Institute and Archives of the Czech Academy of Sciences (research scheme RVO: 67985921) and by the post-doctoral research grant provided by the Czech Science Foundation (No. P410/10/P316). This post-doctoral research grant also made the English translation of the whole book possible. The extraordinarily generous one-year fellowship at the Imre Kertész Kolleg, University of Jena, allowed me to revise and extend the English version. I would like to thank its directors, staff, and fellows for a wonderful intellectual environment. I would also like to thank Berghahn Books for its support, and I extend my gratitude to the three anonymous reviewers whose constructive criticism vastly improved my writing. Lastly, but certainly not least, I thank Caroline Kovtun. Without her exceptional skills as a translator and her infinite patience, English-speaking readers would not be holding this book in their hands.

I dedicate this book to my wife Markéta and our second son Karel, who was born during the very final stage of the preparation of this book. He interrupted me much less than he could have.

Rudolf Kučera
Prague, November 2015

Introduction

From August 2 to 8, 1917 a strike paralyzed production in one of the most important Prague industrial plants—the Ringhoffer railway coach company. Workers refused to work and demanded that their wages and food rations be raised. However, some of them stopped working for other reasons. One of the workers, Josef Plavec, physically collapsed and claimed he could not continue working, not because he wanted to take part in the strike, but simply because of total exhaustion. According to him, the food he was provided was so insufficient that he had depleted his physical strength and could no longer manage his workload. Plavec was apprehended and swiftly brought before a court. There, he repeated his defense, claiming that he could not resume working, not because of his support for the strike, but due to total bodily exhaustion. During his trial, the judge accepted his argument and the court recognized that Plavec could not be prosecuted for taking part in the strike.

However, this did not mean that he was innocent. According to official Austrian regulations, each worker in the militarized industry was guaranteed a scientifically calculated amount of food to provide him with sufficient energy to conduct his or her work. Workloads were measured and used to determine the right amount of calories the workers would receive through food rations. Although the Ringhoffer factory deviated from the established norms and their food rations were slightly lower than those prescribed by the state authorities, the workers were still getting enough calories to conduct their work, albeit with lesser intensity. In the court's opinion, collapsing and being unable to work were thus very unlikely and Plavec's actions were no different from sabotage. His refusal to resume working when his superiors officially demanded it could not be due to bodily exhaustion, but to the intention to harm the Austrian war effort. He was thus sentenced to three years of hard labor and it was only the fact that he did not stop working in order to strike that saved him from a much harsher punishment. According to the law, organized refusals to work carried sentences of up to twenty-five years of imprisonment or, in the most severe cases, death.[1]

The story of Josef Plavec illuminates not just the draconian practices of the wartime Austrian justice system, which stripped many inhabitants of Austria-Hungary of their prewar civil rights and transformed them into mere tools of production for the wartime economy. More importantly, it points to a comprehensive reshaping of the Austrian wartime hinterland driven by pervasive practices of planning and rationing.[2] The scope of the wartime conflict quickly overcame original expectations and caused the entire population previously inconceivable problems, the solution to which often required trying completely untested forms of internal organization.[3] Mobilization for war generated unprecedented pressure for the total and timely reconstruction of the whole economic and social system of the monarchy, allowing little space for the thorough consideration of alternatives.[4] The organization of wartime production and consumption thus had to make do with a mix of foreign, mostly German experiences and prewar, rather theoretical reflections. Although this planning and rationing took place during wartime, its basic contours did not differ from European Enlightenment principles of social planning. The systematic effort to impose a rational order based on scientific knowledge and unlimited human possibilities that would be able to completely transform the world turned the society of the Habsburg Monarchy into a laboratory, in which it was possible to conduct various social experiments that would fundamentally influence the life of all of its inhabitants.[5] During peacetime, these experiments remained behind the closed doors of scientific laboratories and university classrooms. But within the context of the maximum war effort, which influenced the whole society without exception, the vast academic knowledge gathered in the decades before the war provided the blueprints for reform that radically changed the whole monarchy.

The concept of rations thus represents the modern specter of an all-powerful science, able to make decisions in all social conflicts and allocate to everyone exactly what they need based on objective methods. The basic argument of this book is that the notion of a "rationed life," i.e., the notion of a fully rationalized and organized modern world, where everything had to be clearly determined and the location and the amount had to be scientifically justified, took over Czech workers' lives and helped to constitute the wartime working class. The following pages are freely based on Max Weber's classic thesis, which saw the processes of rationalization as one of the main building blocks of European modernity.[6] However, it also updates the Weberian approach with the recent research on the role of science in modern society, which continues in the mostly Foucauldian philosophical tradition. It sees the development of Western society as a constant acceleration of the disciplining of subjects, which, in a rationalized world, occurs especially through the production of knowledge. Scientific knowledge is therefore not merely an explanation of the world around us. Its discourses also produce power relationships and collective identities that can solidify and reproduce themselves precisely through the authority provided by this knowledge.[7]

The book's main subject is the Czech working class. Given the prominence that labor history played in the state socialist historiography before 1989, we can rely on a huge body of literature that has been able to generate a significant amount of empirical knowledge. Many of the relevant works are referenced directly in the text, but Jan Galandauer's and Zdeňek Kárník's books, which remain the most monumental analyses of the development of the Czech working class during World War I, even several decades after their publication, merit special attention.[8] However, the vast majority of Czech works on the labor question prior to 1989 did not actually concern themselves with workers, but rather with the narrowly partisan history of their primary political representative, the Social Democratic party, or, even more narrowly, with the decisions of its wartime party cadres. For many historians, the implicit equation between the large number of industrial workers and a single political party embodied the factory proletariat's emancipation efforts as well as the vanguard of the Communist Party in interwar Czechoslovakia, which in turn was supposed to make the historical mission to establish a communist utopia come true.[9]

Indeed, during the campaign for universal voting rights between the years 1905 and 1907, the Czech Social Democratic party became the largest party in the Bohemian lands with roughly one hundred thousand registered members. The vast majority of these members were also manual laborers.[10] Although one hundred thousand party members represented an admirable number in the context of the times, even in its prime the Social Democratic party was able to win over only a portion of the industrial working class, which, according to the Austrian authorities, numbered roughly one million people in the Bohemian lands right before World War I.[11]

At the same time, the year 1907 was also the year in which membership in the Social Democratic party peaked. World War I dealt a definite blow to the party organization. The party was paralyzed by wartime conscription and for most of the war the Social Democratic party was loyal to the Austro-Hungarian war effort. The significant delegitimization of the party among the rank and file was a consequence of wartime politics.[12] Already before World War I, but especially during the war, there was a considerable gulf between the Social Democratic party and the majority of the working class in the Bohemian lands. If we want to look more closely at the experience of workers, focusing on the Social Democracy Party during the war will not be very helpful.

If older Czech works on the wartime working class provide information on its political representatives while leaving the workers in the background, in the histories of the whole society, however, the situation is quite different.[13] First and foremost is Ivan Šedivý's synthesis, which is still the most complex work on Czech history in the watershed years of 1914–1918.[14] Its second part in particular provides a complex social historical narrative of Czech history during World War I, and it is a good starting point for cultural analyses of wartime society. The last years have also brought renewed interest in the history of social protest[15] as well as in labor and workers, which had practically disappeared after 1989.[16]

Many foreign works focusing not on the Bohemian lands, but rather on the Habsburg Monarchy as a whole or on some of its other parts, provide a broader context for the Czech case.[17] Many are cited in the individual chapters, but two of them deserve to be mentioned here. One is the more than thirty-five years old, but in many respects unsurpassed, work by Richard Georg Plaschka, Horst Hasselsteiner, and Arnold Suppan.[18] Rich in sources, this analysis is primarily devoted to the Habsburg Monarchy's last year of existence and the wave of protests and violence that accompanied its disintegration. Three decades after its publication, this two-volume history remains a monument that cannot be ignored when researching World War I in Central Europe. Out of the more recent works, Maureen Healy's book on the breakdown of the social consensus in wartime Vienna cannot be overlooked. Healy was able to capture the deepening social trenches within the Austrian metropolis that subsequently led to the total collapse of the city, as well as the various wartime experiences of the capital city's inhabitants depending on their social standing, gender, or language.[19] Maureen Healy's work is thus currently the most visible and topical addition to the study of the cultural history of Austrian wartime society.

The retreat of labor history from its formerly prominent place within American and European historiographies has been accompanied by a significant broadening of methodological perspectives.[20] Western historiography thus not only never abandoned the study of the working class as a group that provides modern industrial work, but never even renounced the concept of the "working class,"[21] which for many readers, particularly in post-communist East-Central Europe, evokes a time when the term played a crucial role in the legitimization of socialist dictatorships. The former Marxist-Weberian understanding of the working class as a group of participants connected by their ability to work, which is their only disposable commodity in the free market and from which their other activities are derived, was questioned from all sides. Research on child labor, or the various stages between gainful work and slavery, refuted, for example, the idea of workers' freedom in the modern labor market.[22] The reorientation of historiographical analysis from the individual to the household has shown that work itself was almost never the only disposable article from which workers derived their existence. Home economics, renting modest lodgings or petty theft, embezzlement, as well as hired work all belonged to the workers' arsenal of strategies for subsistence in the nineteenth and first half of the twentieth century.[23]

Due to the transformation of the basic unit of historiographical analysis, in which the individual worker was replaced by the household, there was a fundamental redefinition of the very concept of work, which was freed from direct monetary payment. Work is thus widely understood as "... any human activity that increases the value of goods or services"[24] and, as such, encompasses not only productive work, which increasingly moved into specially designed workplaces during European industrialization, but also unproductive work, which generally remained limited to the sphere of the household.[25]

Within the debates on the "role of the working class in history" that started during the 1980s, the previous primacy of socio-economic determinants was abandoned in favor of multi-causal interpretations, taking into account not only the "social," but also the various cultural variables with a potential to influence the behavior and organization of historical agents[26] and shape the working class's subjectivity.[27] Contemporary historiography does not understand the working class as a product of the objectively measured processes of modernization, but as a very unstable, changing collective that is influenced by many symbolic factors that, depending on the social context, are able to turn part of an amorphous mass of physically working people into a collective historical agent. Revealing and analyzing these symbols in relation to the wartime working class of the Bohemian lands is also the aim of the following pages, which strive to illuminate Czech workers' experience during the "seminal catastrophe of the 20th century," as World War I is often labeled.[28]

Along with the current scholarship, I do not perceive the working class as an already given, closed and objectively existing collective created from suprapersonal structures of economic production and defined by its position within these structures, its salaries and the standard of living, consumption, etc. provided by these salaries. Instead, this book conceives of the Czech working class during World War I as a project that has never been finished—a phenomenon that is constantly forming and transforming at the intersection between cultural and symbolic practices on the one hand and lived experiences on the other.[29] The effort to describe workers' experiences during wartime is therefore meaningful in this context. The aim is not only to describe the banality of people's lives in a given time period, but to identify the places in space and time where the lived experience intersect with the discourses and symbols of the state. Precisely at these intersections collective identities and demands emerge and can be captured. The main questions of this book are thus on the war's influence on the transformation of an organized working class—its culture and the way active workers understood themselves and their surroundings during the rapid wartime changes.

If we understand politics as a sphere where the collective identities and demands of individual social groups are formulated and articulated, and where these groups subsequently clash with the state or each other, we can see that society in wartime Austria-Hungary was politicized at every level, even though, for most of the war, it had neither parliamentary politics nor liberal rights.[30] Under conditions of acute material shortage, the enormous strain on wartime production, and rising social tensions, the dynamic regrouping of social hierarchies occurred more often than ever. New social collectives were created that formulated new demands on each other or on the state. The inhabitants of malnourished towns felt cheated by the agricultural countryside; German-speaking citizens of the monarchy accused their Czech counterparts of insufficient wartime loyalty; Czechs and other non-German ethnic groups felt oppressed in every way; many women accused the male-dominated political system of the monarchy of using them for hard wartime labor but denying them basic civil rights. The majority of the increasingly impoverished inhabitants of the whole country observed with growing bitterness the enormous profits of a narrow number of businessmen who were able to get enormously rich off the wartime economy. All of these groups then turned to the state to acknowledge their demands and solve their problems. In the end, the inability to satisfy these demands brought about the total collapse of the basic social solidarity of wartime Austrian society and, with it, the disintegration of the entire Austro-Hungarian Empire. Thus Austro-Hungarian society was actually more political than ever, and its workers were one of the central building blocks of the Austrian wartime effort, playing a central part in "depoliticized politics." The basic perspective of this book hinges precisely on the initially chaotic fields of mutually intersecting group identities and their demands and collective actions. In the following pages, I understand the "politics of the working class" as those spheres of the Czech workers' experience in which collective identities and collective demands were created, defining the organized workers' collective inwardly and outwardly. My observations of the "politics of the working class" thus led me to divide the book into four main chapters.

The first chapter is devoted to the "politics of food." The question of the distribution and consumption of food was one of the most visible blows to the prewar workers' collective and represented the most serious challenge to the basic survival of all workers. Access to food and its consumption during wartime scarcity was a prominent stage upon which social dividing lines were created and manifested. At the same time, however, the food question became the most important issue of the all-encompassing rationing system that was implemented by the state. Demands for various foods and their acceptance or rejection to a great extent stemmed from the primacy of modern science as a universal source of advice on the organization of life and science, then played a significant role in the wartime "politics of food." Therefore, the chapter analyzes the development of this science as well as its influence on the transformations of workers' lives.

The second chapter is devoted to the transformation of industrial labor as another central factor in the collective self-identification of the working class. Physical labor was one of the basic defining components of an organized working class in the prewar years, and the drastic changes that it went through between the years 1914 and 1918 also significantly influenced the workers' collective. Here, too, several scientific fields held a dominant position, claiming to know universal truths about what constitutes human labor and how, when and where it should be conducted. The chapter focuses not only on these scientific fields, but also on the blending of this knowledge with Austro-Hungarian political power and with the world of hundreds of thousands of workers in the wartime industry.

The third chapter switches perspective to the significantly changed gender composition of the working class. It focuses on the disruption of the prewar male hegemony in the public space of the Habsburg Monarchy as well as within families and at the workplace. Although gender is a sphere in which modern scientific knowledge did not play such a defining role during wartime, even here we can detect its influence on several significant developments in the gender make-up of the working class. The massive influx of women into the wartime industry and the disintegration of the construct of male public authority are also addressed.

The last chapter focuses on the forms of the workers' protests. The mutual interconnection of all the previous "politics" is most obviously revealed precisely in the phenomenon of the wartime workers' protest, because it almost always arose when problems with food distribution, the massive reorganization of industrial labor and radical changes in the gender composition of the organized workers' collective were combined. The changes in the shape of the wartime protests and the composition of the protest groups offer an insight into the collective actions of the newly created working class and its limits. Such limits manifested themselves every time the working-class protest was not able to integrate a greater number of industrial workers.

Although the names of the four chapters may lend the impression that each one is reserved solely for one sphere of wartime "politics," this is not the case. The questions of wartime consumption cannot be separated from the problems connected with industrial labor. The wartime gender diversification of the working class also took place in close connection with the sphere of labor as well as that of consumption, and the wartime workers' protest is connected to the questions of gender as well as those of labor and food. The individual topics run through all of the chapters, but one topic dominates each of them. All of the chapters together attempt to paint a portrait not only of workers' lives in the Bohemian lands during wartime, but also of their contacts with scientific and state authorities and with the other citizens of wartime Austria-Hungary. The central question, however, remains—how did these contacts influence the working class's self-identification and how did they contribute to the creation of the wartime working class as a collective historical agent, or, on the contrary, how did they prevent this from happening?

Notes

1. *Kovodělník. Orgán svazu dělnictva zaměstnaného výrobou a zpracováním kovů a drahokovů v Rakousku* 35 (August 8, 1917): 140–41.
2. I. Šedivý, *Češi, české země a velká válka 1914–1918* (Prague, 2001), 217–18. For the concept of planning and possible methods of its historicization, see H. Klages, "Planung—Entwicklung- Entscheidung: Wird die Geschichte herstellbar?," *Historische Zeitschrift* 226 (1978): 529–46. Thomas Etzemüller, "Social Engineering als Verhaltenslehre des kühlen Kopfes. Eine einleitende Skizze," in *Die Ordnung der Moderne. Social Engineering im 20. Jahrhundert*, ed. Thomas Etzemüller (Bielefeld, 2009), 11–40.
3. For the very outbreak of the war and the debates on what caused it, see C. Clark, *The Sleepwalkers: How Europe Went to War in 1914* (New York, 2012). The most recent synthetic works on World War I include J. Leonhard, *Die Büchse der Pandora. Geschichte des Ersten Weltrkieges* (München, 2014); H. Münkler, *Der Große Krieg. Die Welt 1914–1918* (Berlin, 2013); A. Kramer, *The Dynamics of Destruction: Culture and Mass Killing in the First World War* (Oxford, 2008). For a recent research overview, see A. Kramer, "Recent Historiography of the First World War (Part I)," *The Journal of Modern European History* 12, no. 1 (2014): 5–27; A. Kramer, "Recent Historiography of the First World War (Part II)," *The Journal of Modern European History* 12, no. 2 (2014): 155–74.

4. On mobilization during World War I, see H. Perry, *Recycling the Disabled: Army, Medicine and Modernity in WWI Germany* (Manchester, 2014); R. Chickering and S. Förster, eds., *Great War, Total War, Combat and Mobilization on the Western Front, 1914–1918* (Cambridge, 2006); J. Horne, ed., *State, Society, and Mobilization in Europe during the First World War* (Cambridge, 1997); Michael Geyer, "The Militarization of Europe, 1914–1945," in *The Militarization of the Western World*, ed. J. Gillis (New Brunswick, 1989), 74–79.
5. On the concept of planning, see, for example, D. van Laak, "Planung. Geschichte und Gegenwart des Vorgriffs auf die Zukunft," *Geschichte und Gesellschaft* 34, Heft 3 (2008): 305–26.
6. M. Weber, *Die protestantische Etik und der Geist des Kapitalismus* (Tübingen, 1934); P. Wagner, *A Sociology of Modernity: Liberty and Discipline* (London and New York, 1994). From the recent writings, see, for example, K. Brückweh, D. Schumann, R. F. Wetzell and B. Ziemann, eds., *Engineering Society: The Role of the Human and Social Sciences in Modern Societies, 1880–1980* (Basingstoke, 2012); T. Mergel and C. Reinecke, eds., *Das Soziale ordnen. Sozialwissenschaften und gesellschaftliche Ungleichheit im 20. Jahrhundert* (Frankfurt am Main, 2012).
7. Out of Foucault's large oeuvre, see, for example, M. Foucault, *Power/Knowledge: Selected Interviews and Other Writings, 1972–1977* (New York, 1980).
8. J. Galandauer, *Bohumír Šmeral 1914–1941* (Prague, 1986); Z. Kárník, *Habsburk, Masaryk či Šmeral. Socialisté na rozcestí* (Prague, 1996).
9. For an overview of the Czech historiography of the workers' movement before 1989, see Jana Englová, "Dělnictvo jako subjekt a objekt historického bádání," in *Problematika dělnictva v 19. a 20. století. Bilance a výhledy studia*, eds. S. Knob and T. Rucki (Ostrava, 2011), 34–39; J. Matějček, "Dělnické hnutí v Českých zemích do roku 1914. Emancipace dělnictva, nebo hegemonie proletariátu? Pokus o objektivní hodnocení vývoje hnutí i stavu výzkumu," *Studie k sociálním dějinám* 2 (1998): 153–95.
10. O. Urban, *Česká společnost 1848–1918* (Prague, 1982), 557. For the broader Austrian context, see further Wolfgang Maderthaner, "Die Entstehung einer demokratischen Massenpartei: Sozialdemokratische Organisation von 1889 bis 1918," in *Die Organisation der österreichischen Sozialdemokratie 1889–1995*, eds. W. Maderthaner and W. C. Müller (Vienna, 1996), 21–92.
11. *Bericht der k.k. Gewerbeinspektoren über Ihre Amtstätigkeit im Jahre 1914* (Vienna, 1915), XXXIX-XLVII. On prewar Social Democracy in the Bohemian lands, see, substantially, L. Fasora, *Dělník a měšťan. Vývoj jejich vzájemných vztahů na příkladu šesti moravských měst 1870–1914* (Brno, 2010).
12. For the latest work on the socialist movement in the southern part of the Habsburg Monarchy, see M. Cattaruzza, *Sozialisten an der Adria. Plurinationale Arbeiterbewegung in der Habsburgermonarchie* (Berlin, 2011).
13. For Czech interwar historiography of World War I, see Martin Zückert, "Der erste Weltkrieg in der tschechischen Geschichtsschreibung 1918–1938," in Geschichtsschreibung zu den böhmischen Ländern im 20. Jahrhundert, eds. C. Brenner, K. E. Franzen, P. Haslinger and R. Luft (München, 2006), 61–75.
14. Šedivý, *Češi*. For older Czech works, see Z. Jindra, *První světová válka* (Prague, 1984).

15. P. Heumos, "Kartoffeln her oder es gibt eine Revolution. Hungerkrawalle, Streiks und Massenproteste in den böhmischen Ländern 1914–1918," *Slezský sborník* 97, no. 2 (1999): 81–104.
16. Pavel Kolář and Michal Kopeček, "A Difficult Quest for New Paradigms: Czech Historiography After 1989," in *Narratives Unbound: Historical Studies in Post-Communist Eastern Europe,* eds. S. Antohi, B. Trencsényi and P. Apor (Budapest and New York 2006), 198. Out of the newer works, see especially L. Fasora, *Dělník a měšťan. Vývoj jejich vzájemných vztahů na příkladu šesti moravských měst 1870–1914* (Brno, 2010); S. Holubec, *Lidé periferie. Sociální postavení a každodennost pražského dělnictva v meziválečné době* (Plzeň, 2009); J. Gelnarová, "Matka Praha a dcery její. Diskuse o ženském volebním právu do obce pražské v občanském a dělnickém ženském hnutí mezi lety 1906 a 1909," *Střed/Centre* 2 (2011): 34–58; M. Jemelka, *Na Šalomouně: společnost a každodenní život v největší moravskoostravské hornické kolonii (1870–1950)* (Ostrava, 2008).
17. For a comprehensive overview of the historiography on the late Habsburg Monarchy, see most recently J. Deak, "The Great War and the Forgotten Realm: The Habsburg Monarchy and the First World War," *Journal of Modern History* 86, no. 2 (June 2014): 336–80. For a broader insight into the recent scholarship on World War I, see, for example, J. Bürgschwentner, M. Egger and G. Barth-Scalmani, eds., *Other Fronts, Other Wars? First World War Studies on the Eve of the Centennial* (Leiden and Boston, 2014); E. Krivanec, *Kriegsbühnen: Theater im Ersten Weltkrieg. Berlin, Lissabon, Paris und Wien* (Bielefeld, 2012); S. Goebel and D. Keene, eds., *Cities into Battlefields: Metropolitan Scenarios, Experiences and Commemorations of Total War* (Farnham, 2011); F. Altenhöner, *Kommunikation und Kontrolle: Gerüchte und städtische Öffentlichkeiten in Berlin und London 1914/1918* (München, 2008); C. Acton, *Grief in Wartime: Private Pain, Public Discourse* (London, 2007).
18. R. G. Plaschka, H. Hasselsteiner, and A. Suppan, *Innere Front. Militärassistenz, Widerstand und Umsturz in der Donaumonarchie 1918,* Band I.–II. (München, 1974).
19. Healy, *Vienna.* On the German case, see B. J. Davis, *Home Fires Burning: Food, Politics and Everyday Life in World War I Berlin* (Chapel Hill, NC and London, 2000).
20. For the Western European historiography of labor and the working class, see, for example, S. Berger, "Die europäische Arbeiterbewegung und ihre Historiker. Wandlungen und Ausblicke," *Jahrbuch für europäische Geschichte* 6 (2005): 151–82; T. Welskopp, "Von der Verhinderten Heldengeschichte des Proletariats zur Vergleichenden Sozialgeschichte der Arbeiterschaft—Perspektiven der Arbeitergeschichtsschreibung in den 1990er Jahren," *Zeitschrift für Sozialgeschichte des 20. und 21. Jahrhunderts* 3 (1993): 34–53. Further, for example: M. van der Linden, *Workers of the World: Essays Towards a Global Labor History* (Leiden–Boston, 2008); L. Heerma van Voss and M. van der Linden, eds., *Class and Other Identities: Gender, Religion and Ethnicity in the Writing of European Labor History* (New York, 2002).
21. G. Eley and K. Nield, "Farewell to the Working Class?" *International Labor and*

Working Class History 5 (Spring 2000): 1–30; J. W. Scott, "The Class We Lost?" *International Labor and Working Class History* 5 (Spring 2000): 69–75.
22. M. Rahikainen, *Centuries of Child Labor: European Experiences from Seventeenth to the Twentieth Centuries* (Adlershot, 2004); T. Brass and M. van der Linden, eds. *Free and Unfree Labor: The Debate Continues* (Bern and Frankfurt am Main, 1997).
23. Van der Linden, *Workers*, 26–27.
24. C. Tilly and C. Tilly, *Work under Capitalism* (Oxford, 1997), 22.
25. M. Burawoy, "The Anthropology of Work," *Annual Review of Anthropology* 8 (1979): 231–66.
26. For a good introduction to the "cultural wars," as the discussions on the roles of the material and cultural determinants of the actions of historical participants are sometimes called, and for a basic bibliography, see G. Eley and K. Nield, *The Future of Class in History: What's Left of the Social?* (Ann Arbor, MI, 2007); L. Heerma van Voss and M. van der Linden, eds., *Class and Other Identities: Gender, Religion and Ethnicity in the Writing of European Labor History* (New York, 2002); L. Frader, "Dissent Over Discourse: Labor History, Gender and the Linguistic Turn," *History and Theory* 34 (1995): 213–30.
27. For the latest, see, for example, L. Bluma and K. Uhl, Hg., *Kontrollierte Arbeit—Disziplinierte Körper? Zur Sozial- und Kulturgeschichte der Industriearbeit im 19. und 20. Jahrhundert* (Bielefeld, 2012); P. A. Custer, "Refiguring Jemima: Gender, Work and Politics in Lancashire 1770–1820," *Past & Present*, no. 195 (May 2007): 126–58.
28. G. F. Kennan, *The Decline of Bismarck's European Order: Franco-Russian Relations 1875–1890* (Princeton, NJ, 1979), 6. Further, see: W. J. Mommsen, *Die Urkatastrophe Deutschlands. Der Erste Weltkrieg 1914–1918* (Stuttgart, 2002); Hans Joas, "Kontingenzbewusstsein: der Erste Weltkrieg und der Bruch im Zeitbewusstsein der Moderne," in *Aggression und Katharsis: der Erste Weltkrieg im Diskurs der Moderne*, eds. P. Ernst, S. Haring, and W. Suppanz (Vienna, 2004), 43–56.
29. Eley and Nield, *The Future*, 171.
30. For this understanding of politics, see, for example, the seminal works L. Hunt, *Politics, Culture, and Class in the French Revolution* (Berkeley and London, 1984); J. Vernon, *Politics and the People: A Study in English Political Culture c. 1815–1867* (Cambridge, 1993).

Chapter 1

Rationed Satiety: The Politics of Food

"In this war one thing would delight me,
O Dumpling, thy breadcrumbs enslave me!
I dream of thee, Dumpling dear, nightly,
Floating in meat sauce and gravy"
— Karl Kraus, *Die letzten Tage der Menschheit* [1]

An Unexpected Visit

The afternoon of Monday, November 29, 1916 began just like any other in the villa of textile factory owner Ludvík Jelínek in the Prague neighborhood of Libeň. Jelínek stayed home after breakfast and was likely working while his wife was overseeing the preparation of lunch. His maid lit a fire in the stove and placed two raw goose thighs and half a kilogram of lard on the table, which was to be the basis of the day's lunch. But just before she could place the thighs into a hot roasting pan, the idyllic afternoon was disrupted by an urgent knock on the door. Before anyone in the family could react, the quiet Monday suddenly turned into a nightmare. Policemen burst into the villa and without hesitation went straight into the kitchen. Everything happened so quickly that nobody in the family or any of the servants could put up any resistance. The two goose thighs, the roasting pan and the melted lard were seized as evidence,[2] and Ludvík Jelínek was served with a summons to the local police station, where he was requested to give a statement regarding the misdemeanor charge of violating the imperial decrees no. 218/1916 from July 14, 1916 and no. 285/1916 from September 1, 1916 on meatless days. Both of these decrees forbade the sale and consumption of meat and meat products on Mondays, Wednesdays, and Fridays. This applied to butcher shops, restaurants and public kitchens as well as to individual households.[3] The punishment for violation was a fine of up to 5,000 crowns or imprisonment for up to half a year.[4]

Due to the small amount of meat seized and its private usage, the Prague municipal office punished Jelínek's forbidden lunch with a fine of 500 crowns. Ludvík Jelínek, however, fought back. Through his attorney, Richard Singer, he filed a suit against the municipal office's sentence with the administrative court, in which he demanded a revision of the fine. Singer argued that Jelínek did in fact violate the decree, but that the forbidden act of consuming meat at the time of the police raid was merely attempted, for the goose thighs, as the official report states, were lying on the table and had not yet been eaten. Furthermore, according to Singer, the evidence could not prove that Jelínek had not intended to eat both goose thighs on the following day, when consuming meat was permissible by law. The fine should therefore either be rescinded or, at the very least, reduced.

Unfortunately for Jelínek, however, the court did not agree with this legal argument. Although the goose thighs were not yet eaten at the time of the police raid, the judge deemed the fact that the hot roasting pan was greased with lard irrefutable proof that, were it not for the timely intervention of the police, meat consumption would certainly have taken place. Also, the use of lard for cooking was the legal basis for the fine, for according to the judge the consumption of meat or meat products was not understood merely as the act of eating, but also of their further usage in the preparation of meals.[5]

In June 1917, the court upheld Jelínek's fine. The famous Prague glove and fur manufacturer, however, continued to see 500 crowns as a rather inflated price to pay for two goose thighs, especially when they had been seized by the relevant authorities. Despite the court's decision, he stubbornly refused to pay the fine even after many reminders, giving in only when the Prague municipal office handed down an order of repossession.[6]

A week after the unpleasant visit to Jelínek's Libeň villa on that Monday morning, another impatient knock was heard, this time at the doors of the Prague apartments where the workers František Žižka and František Chlupatý lived. Both spent Monday morning at work and so the act of standing by during a humiliating police raid was left to their wives, in Anna Žižková's case accompanied by the incessant crying of two small and hungry children. Here, too, sufficient incriminating evidence was collected. Roughly 60 grams of beef, which Žofie Chlupatá was warming up for lunch, was convincing enough proof for the police, along with the roughly 125 grams of horse meat that Anna Žižková was preparing for her family. The meat and its juices and sauce were seized, and both women were summoned to give their statements.[7]

Žofie Chlupatá defended herself by citing her abject poverty, which official Prague reports confirmed. With a husband working as an assistant galvanizer, they could not afford to get anything more to eat than what their extended family could provide for them due to the rising prices of all necessities. The seized meat had been brought by her sister the day before the police inspection, while the roughly 60 grams of evidence was, according to the official statement, "hard and of a bad quality."[8]

As a mother of two children, Anna Žižková received government alimentary benefits, but even her family's situation did not allow for much luxury. The 125 grams of horse meat confiscated during the morning inspection was supposed to feed her, her husband, who was providing for his family by working as mechanical worker, and their two children until the next day, when the family was to receive more ration cards.[9]

None of the women could afford legal assistance, and their financial situation ruled out the possibility of imposing a fine. The only thing that the Prague municipal office could do in this case was to sentence them to jail. The length of the sentence is not mentioned in the sources, nor is the fate of the children of the Žižka family, who had nobody else to take care of them.

The Dream of the Perfect Fuel

The unfortunate episodes in the lives of the families of factory-owner Jelínek and the workers Chlupatý and Žižka show just how significant the regulation of food consumption was in the everyday life of the Habsburg Monarchy during World War I. With no regard to differences in social standing and status, three families were subjected to the same harsh disciplinary regime under conditions of dwindling resources. The merciless system of control—in this case the control of meat consumption—did not stop at the doors of private apartments and houses, thus perfectly illustrating the new, state-regulated social order that had emerged in August 1914.[10]

As the war continued, worsening food shortages began to dominate life in the monarchy, eventually paralyzing the whole war effort. This chapter will focus on this aspect, that is, on the questions connected with obtaining and consuming basic foods, which greatly influenced everyday life in the Bohemian lands. The first part analyzes the changing relationship between the consumption of food and manual labor. The focus will be on natural science discourse, which largely determined the understanding of the nature of food and how its consumption could contribute to defeating the enemy. This discourse also prescribed how the food requirements of individual social groups were to be satisfied during the war years. The second part examines the influence of this discourse on the wartime "politics of food,"[11] that is, the public articulation and satisfaction of (or overlooking of) demands for certain foods, and the influence of these politics on the transformation of the working class. It was precisely the politics of food that upset the prewar social divisions and caused both the establishment and subsequent disintegration of the wartime social order. And, as the urban industrial working class was the most affected by the changed socio-economic transformations of the wartime state, it played a central role in the final collapse of the monarchy in the fall of 1918.

In order to understand the relationship between nutrition and wartime industrial production it is necessary to look back to the prewar decades. Prior to 1914, the biblical adage "he who does not work may not eat"[12] underpinned the liberal economic order in most of industrialized Europe. The more or less unrestricted market oversaw food production and distribution, particularly in urban centers, and access to food was thus tied to the salaries earned through paid work. On the other hand, the nexus between work and nutrition underwent a dynamic change during the second half of the nineteenth century, resulting in a totally new conceptual framework that would eventually determine everyday wartime reality in the Bohemian lands.

The sea change in understanding what manual labor actually was occurred in the years 1847 and 1865, when Hermann von Helmholtz and Rudolf Clausius published the first two laws of thermodynamics about wireless energy transfer in isolated systems and loss of energy (entropy) occurring whenever energy is transferred from a warmer object to a colder one.[13] Together, these discoveries created the basic conceptual framework that structured thinking about human labor in the second half of the nineteenth century. The principle of energy transfer and the possibility of conserving or losing energy meant that labor was conceived as a process of energy transformation: input energy was transformed into labor output. Labor thus ceased to be imagined in categories of Christian morality and was integrated into the discourse of modern physics. It became an objectively defined quantity measurable in precisely determined units—joules.[14] The word "labor" in the second half of the nineteenth century thus became a general term, first in German and French and later in other languages, denoting any creative activity that entailed the output of energy regardless of whether this energy was expended by inanimate machines or living humans.[15]

This blending of human and machine labor created a new language to describe the workings of the human body. While the understanding of bodily functions until the mid nineteenth century was based on the dominant metaphor of the human body as a mechanical machine, by the end of the nineteenth century the new thermodynamic discoveries of von Helmholtz and Clausius brought about a gradual change in the ideas about what the human body really was and how it worked. The research on the human body and how it functions when it receives and transforms energy amassed evidence that the human body, unlike the classic mechanical manual drive machines, is active even when it is not working, because even in an idle state it releases an appreciable amount of thermal energy. Thus, instead of the dominant metaphor of the human body as a mechanical machine, the conception of the human body as a modern motor—a motor driven by various fuels—slowly gained ground. Like the human body, such motors release energy even in an idle state, i.e., on standby, when not engaged in any work.[16] In the second half of the nineteenth century, practically all of industrial Europe was using a new scientific language of labor that did not differentiate between human labor and the labor of the newly installed gasoline or electric motors. At the end of the nineteenth century, the human body was no longer seen in the imaginations of physiologists, biologists, biomechanics, or diet specialists as an analogy of the industrial motor, but as one of its types. As one of the founders of modern physiology, Carl Ludwig, said: "The steam engine has done a greater service to our science than to any other."[17]

Right before the start of World War I, this new paradigm of the natural sciences was rooted in a hybridization of mechanical machines and human bodies that scientific language could no longer separate in many cases.[18] Synthetically, this defining framework for understanding the human body was summarized by probably the most prominent representative of continental biomechanics, the Parisian professor Jules Amar. His book symptomatically titled *The Human Motor and the Scientific Foundations of Labor* quickly became a classic in the field and the basic compendium of scientific research on industrial labor. Amar argued that just like gasoline or electrical motors, the human motor is also subject to the natural laws on the transfer of energy as formulated by Helmholtz and Clausius. According to Amar, in the case of the human motor, the release of energy was detectable especially in the form of the heat radiating from a human body and the workings of the muscles, while its intake could be calculated based on the chemical composition of the consumed nutrition. The most urgent task of the human sciences was thus to discover a method of energy intake and release that would optimize the body's capacity for work and achieve a maximum increase in work output. Finding the best kind of food, along with appropriate adjustments to working processes in sweatshops, promised to eliminate ineffective moments in factory production, which stemmed above all from bodily exhaustion. Just like suitable fuel and maintenance for inanimate motors would lead to their practically limitless usage, coming up with more suitable food and a better organization of the working day would eliminate ineffective work, i.e., bring about a state in which it would be possible to work without any signs of fatigue.[19]

Amar's theses constituted a summary of the scientific concepts accumulated in the two or three decades prior to World War I. Between 1900 and 1914, the scientific discourse that combined diet, biomechanical or labor-scientific knowledge developed a very complex conceptual framework for the understanding of human nutrition and its relationship to manual labor. The experiments and publications of Russel H. Chittenden, Mikkel Hindhede, or Max Josef von Pettenkoffer, all of whom studied the significance of food for human labor, created a massive body of work in the decade before 1914. Seeing food in terms of the physical categories of transforming energy into labor quickly became the standard view all over Europe. As Anson Rabinbach noted already in 1992, a result of this new scientific understanding of the working human body was its subordination to the discourse of maximum effectiveness that had previously been dominated by the thinking about inanimate machines. Human labor was thus taken out of its social and cultural context and research on it was concentrated on maximizing its output and efficiency. The primary site of industrial work ceased to be the social or cultural context of labor, its location, or those who performed it. Instead, the wholly decontextualized working body became the primary site of industrial production shortly before World War I.[20]

Departing from the objective laws of thermodynamics, attention was placed not only on the conditions of the release of energy, i.e., the work output itself, the organization of work, etc., but also on the questions of its intake, i.e., supplying working bodies with enough suitable nutrition. The metaphor of the human body as a motor was connected with a coherent set of ideas about how a human body functions and what the intake of nutrition means for this functioning. As Jakob Tanner writes in his work on the development of modern nutrition: "A new understanding of human beings came about that logically integrated their nutrition, labor and social utility. The first law of thermodynamics on conserving energy provided a theoretical foundation for research on how the energy chemically contained in food transformed through metabolic processes into kinetic energy (warmth and labor). The human body was understood as a matrix of input and output. On the input side, nutrition provided the necessary energy from which the body produced mechanical power on the output side."[21]

Around 1900, a number of scientific fields therefore focused on the energy effectiveness of the individual components of nutrition and their usage by the human motor. Subordinating the questions of manual labor to the demands for maximum rationalization and effectiveness soon added a scientifically rationalized dimension to the well-known biblical expression of he who works must also eat. The fundamental questions, however, were what, when, and how.[22]

Already before the outbreak of World War I, European diet specialists and physiologists were researching the maximum rationalization of nutrition, which was understood as the largest possible increase in work output using the least amount of food. This research was conducted at a time when modern science was already fully established not only as the source of "objective" knowledge, valid within individual fields, but also when the very person of the scientist was seen as a respected source of practical instructions for the organization of everyday life.[23] As Philipp Sarasin notes, at the end of the nineteenth century the natural sciences in particular replaced the authority of Christian dogma with their emphasis on the fact that they were discovering the laws and inevitabilities of nature: "Universally valid natural laws replaced the previous laws of God.… Whether it was Newtonian mechanics, the laws of gravity, Darwinian selection, thermodynamics or, later, the laws of genetics, the only deciding factor was the empty signifier of the 'law,' i.e. the regularity, inevitability and naturalness that connects the progress of the world with the life of the human body.… Its personification was the fatherly figure of the scholar, whose name guaranteed that lay people would submit to these symbolics."[24]

In Europe at the end of the nineteenth century, this transformation of the social status of science and its human representatives in particular established the figure of the scientist as the provider of universal instructions, the validity of which did not need to be further justified. Science itself had the capacity to solve all problems, including those connected to human nutrition. The fall of 1914 was a fundamental turning point, when the newly arisen critical supply situation in Germany, and even more so in Austria-Hungary, provided the ideal context for the rapid implementation of the prewar scientific findings.

The first framework for this implementation was created at the start of the fall of 1914. In Austria-Hungary, a series of imperial or government decrees were passed, laying the groundwork for a complex government rationalized system of distributing and consuming food, which ultimately dominated the entire Habsburg Monarchy in the last four years of its existence. On October 10, 1914, the Austrian government was empowered by decree no. 274/1914 to supply the population with food in the public interest, giving it free reign to pass any other legislation.[25] The first to be regulated was the flour market, when on October 31, 1914 regulation no. 301/1914 mandated that bread be made with mixed flour, in which the content of wheat or rye flour would not exceed 70 percent.[26] In December 1914, the maximum price of flour was established. On February 26, 1915, regulation no. 46/1915 drastically limited the usage of wheat and potatoes in alcohol distilleries,[27] and on March 20, 1915, decree no. 70/1915 gave the provincial administrations the authority to ban the production of smaller baked goods.[28] During the year 1915, this developing system of limiting consumption and controlling prices was gradually supplemented with a central system of distribution. In April 1915, a ration system was put in place for flour and bread. One year later, a similar system was established for the distribution of meat, the maximum price of which was set in mid 1916. Potatoes and other basic foods were rationed as well, and Austria-Hungary thus became the main pioneer of the ration system among the wartime states.[29]

The state thus took the distribution of food into its own hands, but without canceling the role of the market. Food continued to be officially sold for cash, which was, however, supplemented by various cards that authorized people to purchase specific kinds of foods. The state guaranteed each citizen the right to receive precise rations of meat, milk, oil, bread, etc., but this right could only be exercised by paying the respective market price. Therefore, the ration system was not able to eliminate the consequences of the supply crisis, especially for the urban classes dependent on a regular salary, for the value of their salaries was quickly devaluated by the rapidly increasing wartime inflation, when the average income decreased significantly in comparison with the prewar period.[30]

In contrast to the previous period of peace, for the entire duration of World War I, the constantly adjusted centralized system of food distribution and rationalization was the most striking intervention into the social order, and it drastically influenced the wartime experience of all the inhabitants of the Habsburg Monarchy. As Maureen Healy writes in her seminal work on the collapse of the social consensus in wartime Vienna: "World War I introduced a novel and important variable into the tangled web of Viennese social identities: one's access to or distance from food."[31] There is no reason to think that the situation in the Bohemian lands during the war was qualitatively different from the Viennese experience. Despite several regional differences, the Bohemian lands also went through an unexpected supply crisis during World War I that tested the limits of imaginable poverty further with every year.

Due to the dwindling supply of food and the growing cracks in the distribution system, it became increasingly difficult for the Austrian authorities to convince the population of the meaningfulness of the great sacrifices that were being officially and universally demanded. While the wartime propaganda churned out appeals stating that the most important task for all social classes was to tighten their belts and "bear" (*durchhalten*) the burdensome situation for their motherland and the throne until the triumphant end, the language of universal sacrifice was regularly undermined by daily confrontation with the ubiquitous scarcity of basic human necessities.[32]

During this time, diet specialists and physiologists came down from their laboratories and university departments and significantly contributed to the debates on the wartime supply situation. They did so in the context of the changed social status of the natural sciences and its representatives, who had already become the unchallenged advisors on various aspects of everyday life before the outbreak of the war. A German professor of physiology at the University of Halle, Emil Abderhalden, aptly described wartime alimentation in the Bohemian lands in the introduction to his popularizing work on wartime nutrition: "Great are the performances of science and technology, especially in today's period of war. The moment a problem arises, it is immediately solved."[33]

Precisely according to this logic, the wartime supply problem was one of the issues that could definitely be solved thanks to the primacy of scientific knowledge in the everyday life of the wartime society. As the Prague doctor Antonín Merhaut noted in 1916 in his articles written for the influential Czech liberal daily *Národní Listy*: "Scientific works on human nutrition can today be considered perfect and therefore authoritative for human society and they have long been the basis for determining food rations …, so that there would be precisely the kind of nutrition to suit everyone's needs."[34]

Objective and objectifying science thus decisively entered into the social debates on individual groups' demands for rations of certain foods. However, scientific research was not undertaken with an effort to ensure new taste experiences or contribute to expanding the feeling of fullness. In fact, it was driven by an effort to find the most effective fuel to energize working bodies: "Nowadays the central question is not how many nutritious elements are normally ingested, but rather how many of them are really necessary for sufficient sustenance. It is necessary to examine how many of the ingested nutritious elements are truly consumed and expended in the body. … The elemental and energetic expenditure of everyday life also determines the need for an exact nutritious intake," wrote Professor František Mareš, one of the most prominent Prague physiologists on the Medical Faculty at Charles University, in one of his popular books.[35]

The main criterion that determined the number of food rations for manual laborers was the value of the thermal energy that the human body could acquire and use to maintain its basic functions—the simple running of the human motor—and convert into physical work. Establishing the energy values of the different types of food was not difficult. The caloric values of the three main nutritional components—proteins, carbohydrates, and fats—had been known since 1902, when the Berlin professor of hygiene Max Rubner published his work "*The Laws of Energy Consumption in Nutrition*."[36] Knowing the relevant values and the most renowned personalities that helped to determine them was required of every student of biology or medicine, and this often spread to popularizing works geared towards a wider public.

The Prague professor and director of the Bohemian Institute of Agronomy, Julius Stoklasa, for example, drew on the authority of more than fifteen leading European and American physiologists, such as the Germans Georg Forster and Friedrich Wilhelm Beneke, the Dane Mikkel Hindhede, and the American Russell H. Chittenden, in his justification of the wide-ranging reforms of eating habits. According to him, these scientists determined "the consumption of food necessary for human alimentation in physical-chemical measures. The heat—the caloric effect—of individual nutrients determines their nutritional value. 1g of protein equals 4.1 calories in the human body. 1g of carbohydrates produces the same amount of calories, whereas 1g of fats offers 9.3 calories. An idle body consumes a certain amount of thermal units to maintain its metabolism, the invisible, silent cardiac function, the maintenance of the organism's bodily temperature and similar essential bodily functions."[37]

While the wartime Habsburg Monarchy could rely on the scientific findings shared practically all across the industrial world to establish the energy values of basic foods, when it came to establishing the optimal amount of overall energy that would guarantee the most effective functioning of the human body, the situation was more complicated. Despite the fact that various scientific teams were conducting intense research and results were shared all over Europe, scientists in this field did not reach the same conclusions. The question of what percentage of energy the human motor was really capable of utilizing from the supplied nutrition was contentious. Just like for any other machine, it was hence necessary to establish a rate of loss: "Now we know that no machine created by man can convert all of the energy into work. The largest part of this energy is lost in the form of evaporating heat. Experience has shown that the muscle machine can convert about 20 percent of its total energy intake into work," Emil Abderhalden concluded based on his careful study of the relevant literature and his own research.[38]

Another problematic question was the exact amount of energy that the individual types of labor consumed. Here, the Bohemian lands were influenced by the experimental research of Carl von Voit, a professor of physiology at the University of Munich, who conducted rather extensive research among Bavarian workers already in the 1870s. From his long observation of their work output and food consumption, he deduced the daily amount of calories necessary to keep a human body running depending on specific physical tasks. For a male manual laborer doing a moderately difficult task, he established this value at 2,691 calories; for women it was 2,153.[39] However, other research did not confirm these values, and the calculations of indispensable caloric intakes differed throughout Europe. Thus, for example, the French physiologist Armand Gautier proposed average values of one hundred less calories in his research, while distinguishing even more between the age of the worker and the difficulty of the task. For certain types of physical exertion, he even conceded caloric values that were twice as high as those noted by von Voit and his team.[40]

Although there were disagreements on the questions of the amount of calories needed for certain types of work, the values determined by various experts for an average male laborer were between 2,400 and 2,700 calories per day. (By comparison, the current minimal norm defined by the United Nations is roughly 1,800 calories per person per day; the average Czech has a daily intake of 3,320 calories.)[41] An even larger disproportion emerged during efforts to solve the issue of how to provide this amount of energy for the human body most effectively, or rather what the optimal ratio of the three basic food groups in the consumed food should be. According to von Voit's findings, the average Bavarian worker needed 118 grams of protein, 56 grams of fats, and 500 grams of carbohydrates daily in order to reach the advised 2,691 calories.[42] Similar research in France and England, however, reached vastly different conclusions. According to the well-known findings of Lyon Playfair, English workers consumed 176 grams of protein, 71 grams of fats, and 666 grams of carbohydrates to achieve a similar amount of calories. French workers, according to the previously mentioned Armand Gautier, used 190 grams of protein, 90 grams of fats, and 600 grams of carbohydrates to get the energy they needed for their daily work.[43] So, for example, in terms of fats, the various values differed by more than half.

Although none of the above-mentioned research was unknown in the Austro-Hungarian Monarchy, the local nutritional science was still heavily influenced by the research of the "Munich school" around Carl von Voit. However, almost immediately after the outbreak of the war it became evident that the Austrian supply system could not provide the minimum nutritional values prescribed by von Voit. When the ration system was first put in place, the total amount of energy that the inhabitants of Austrian cities, who were unable to independently provide food for themselves, were to receive was calculated at a mere 1,300 calories per day, and as the war went on this amount decreased to 830 calories. The actual supply was even more modest.[44]

Austro-Hungarian diet specialists therefore repeatedly reevaluated some of the main conclusions of von Voit's school and tried to adapt its main findings to the context of ever dwindling supplies. Due to the tense meat supply situation—in several towns meat had been unavailable since the end of 1916—the central topic was the role of animal proteins.[45] During the years 1915 and 1916, a group of doctors, biologists, and diet specialists from different Prague universities intensively conducted laboratory experiments and kept up on the relevant foreign literature. The result was a very wide public campaign that occurred parallel to the growing food crisis between 1915 and 1916. The goal of this campaign was to change eating habits in the name of victory, and to place the wartime economy, troubled by an oppressive food shortage, on sustainable foundations. Czech newspapers were literally swamped with various tips on how to achieve the most rational nutrition possible to guarantee at least the sustainment of the current work output under conditions of pervasive wartime scarcity.

The basic premise of this public campaign to change eating habits was the refutation of prewar nutritional charts, which, although the numbers of the specific values differed, were all based on the irreplaceable role of meat as the best source of protein.[46] The nexus between the meat and fat-based diet and hard manual labor was deeply ingrained especially in the workers' minds long before the First World War, but for the Habsburg wartime administration the consumption of meat was the biggest burden on the central supply system.[47] Hence, wartime nutritional science began to dismantle deep-rooted ideas on the right diet; it pictured Czechs as eating poorly, with no strategies for improvement, which undermined not only their health, but also their work performance and together with it the war effort of the entire Habsburg monarchy: "We can trace this fact in many adults, who as a result of a bad diet … lose their fitness for their long-term employment…,"[48] scolded the Prague doctor Antonín Merhaut on the pages of the *Národní Listy*.

Traditional tastes were quickly labeled as products of natural, uncivilized instincts and thus irrational. Julius Stoklasa, for instance, argued in one of his educational works on the scientific unsustainability of a traditional, meat-based diet: "Today we know that our protein intake, established by the Munich school around von Voit and Pettenkofer at 118g per day for a working individual, does not correspond to the facts. Already before the war the American Chittenden claimed that only 50-60g of protein per day are necessary to preserve strength and health. The Danish physiologist Hindhede arrived at similar conclusions, documenting that 50g of protein per day were not only enough for regular manual labor, but that gaining weight was also possible with this amount. … 50g of protein per day is perfectly sufficient."[49]

From a scientific standpoint, a meat-based diet thus was not necessary for manual laborers. In the eyes of the middle-class scientists, however, restricting it often collided with popular ignorance and the working class's typical tendency to give in to primitive natural instincts: "Since the majority relies simply on their tastes when selecting food and does not take into account their dietary and physiological aspects, we should not be surprised that nutrition that is aimed solely at satisfying the sense of taste is accompanied by dire consequences for the human organism…,"[50] Stoklasa said about those who refused to relinquish their demands for a meat-based diet.

According to the wartime nutritional science, the habit of connecting hard manual labor to the consumption of meat, however, was not merely unnecessary, but also outright unhealthy: "… protein was equated with meat, so that it seemed that there was no protein without meat and that meat was the only perfect food. But unrestrained meat consumption is harmful to one's health and to the economy,"[51] stated professor František Mareš. In his lectures for various workers' associations, the curator of the Museum of the Kingdom of Bohemia and the famous Czech promoter of modern nutritional science, Josef Kafka, repeatedly cited the experiment conducted by an unspecified French biologist, who fed his dog only beef broth, while letting his other dog starve. The dog that was fed only beef broth allegedly died a few days before the dog that was tortured with hunger. The conclusion to draw from this example was clear—it is healthier to eat nothing at all than to eat meat.[52]

František Mareš, the foremost Czech authority on physiology, went even further in his public speeches: "It became apparent from the research that more protein is consumed than is absolutely necessary, which, by the way, simple everyday experience corroborates. It is undeniable that humans wish for a surplus, especially when it comes to food; the rich can afford this and the poor long for it … A worker with a difficult manual job needs a greater amount of general nutrition and his consumption of protein is greater. He could use sugar to take care of this greater need and with greater benefits. But it is not up to him to decide what food he should eat; this can only be based on scientific research and information."[53] In Mareš's version, the whole campaign did not exist simply to enlighten the general public, but was an authoritative ordinance of what food and when was put into the human machine's receiving end.

Those instructed were denied the liberty to decide what they wanted to, and should, eat. Popularizing the newest physiological and dietary knowledge was grounded in the asymmetry between the instructing scientists and the instructed consumers. This asymmetry arose precisely as a result of the scientification of the questions around food, when authoritative knowledge was removed from traditional popular culture in the decades before the First World War, and was gradually landed behind the closed doors of scientific laboratories.[54]

This asymmetry not only reflected the differences in the access to modern knowledge, but also the social inequality that produced it. The instructors were middle-class men, but their instructions were directed primarily toward the working class, often directly at working-class women, who were seen as the administrators of their families' kitchens. The dissemination of information on rational nutrition was imbued with traditional middle-class stereotypes, imagining the working-class environment as espousing opposite values and falling prey to uncivilized, irrational urges and succumbing with relish to their animal instincts. The traditional middle-class topos connecting hard manual labor with the excessive consumption of alcohol was often cited: "… we must once and for all modify our daily meal, so that we can focus on strengthening our jobs after breakfast and would not need to engage our stomachs in consuming a snack during work, which is often accompanied by beer, wine or, for the working class, hard liquor, so that we would not return after lunch to further work lazily and with a laden stomach, but refreshed and full of energy.", doctor Josef Kafka urged working-class women in his lectures.[55]

The newly accumulated information from nutritional experts provided scientific justification for the reduction of already meager food rations practically every day. Before the war, almost all of the Czech physiologists cited the Munich school around Carl von Voit and its figures of a necessary daily dose of 118 grams of protein, 56 grams of fats and 500 grams of carbohydrates as the greatest authority, but at the beginning of 1916, scientifically supported values that argued for under 100 grams of protein for an average hard manual labor job were published in Czech newspapers.[56] In 1917, Czech nutritional science even announced that, according to the latest research, "… 40–60 grams of protein per day per person are wholly sufficient."[57]

As meat gradually disappeared from the markets and shops of most Czech towns during the second half of the year 1916, the scientific instructions for rational nutrition ceased trying to discredit it from the point of view of its negative impact on health and work output. Instead, there was a search for alternative sources of protein, the consumption of which could be limited according to modern scientific findings, but never fully stopped. Professor František Mareš cited the experiments of the American physiologist Benedict, who did not feed his cows any protein. According to his observations, the cattle began to visibly weaken after about three to four years, which suggested that the long-term lack of proteins could have harmful consequences for people as well.[58] Therefore, it was necessary to find other ways of getting an optimal amount of protein as quickly as possible into the human motors driving the Austrian wartime economy in order to guarantee continued work output.

If the origin of the fuel did not matter for combustion engines, it did not matter for working bodies either. Using this simple comparison, Julius Stoklasa, the most important wartime Czech diet specialist, made several laboratory experiments, in which he searched for the best source of protein. By continuously feeding laboratory dogs mucilage, and then analyzing their muscle activity, he came to the conclusion that mucilage can replace meat as a source of protein for up to 80 percent. According to him, mixing mucilage into food was suitable especially for manual laborers, who badly needed a higher level of protein in their diet.[59]

Similar recommendations based on precise chemical analyses and physiological experiments only underscored how questions of nutrition were taken out of their socio-cultural context and set within the framework of the logics of physics—seeing physical work as a mere mechanical conversion of energy.[60] During the First World War, the relationship between nutrition and physical labor became the subject of the complex discourse of modern science, which outlined a strict disciplinarian regime based on the latest nutritional findings. The traditional diet was disqualified not only for being impossible under the conditions of wartime shortages, but especially for being unhealthy and even unsuitable for people with physically demanding work. This was a strong argument within the context of the wartime mobilization. In a society where physical collapse at work was seen as treason and was punished with imprisonment, anyone who continuously disrespected the scientific findings on rational nutrition risked being accused of undermining the war effort.

The increasingly frequent inclusion of dietary questions in the discourse of physics and dietary science during the war initially had the potential to politically neutralize the food shortage problem. The moment science confirmed that the changed and impoverished wartime dietary regime did not prevent a human body from doing the work expected from it, it was no longer necessary to wage a political debate about food rations. The whole project of rationalizing food was based on the premise that the working class actually shared part of the responsibility for the catastrophic supply situation. Its unsuitable dietary habits, learned before the war, slowed down the optimal usage of resources in the country. The war, the insufficient harvest or the allied naval blockade was not solely to blame for the catastrophic scarcity of food. The intractable workers, who stubbornly refused to submit to objectively known truths, were also held responsible.[61]

The discourse of nutritional science was mobilized as a disciplinary force in Austrian daily life. The formulation and dissemination of newly gained scientific knowledge blended with the government's demand to scientifically justify constantly decreasing food rations, as well as with the growing conflicts within wartime society. Scientific knowledge was deployed to address the rising unrest in Habsburg factories and provided arguments for how to quell it. In a society where capital and labor grew more and more distant from one another every day, one of the strategies for preventing a total collapse was finding objective, and therefore generally shared, norms.

Nutritional science offered precisely such norms, and thus also the opportunity to remove aggressive, class-based language from the communication between the ruling elites on the one hand, and the broader social classes on the other hand, and to base this communication in consensus guaranteed by "objective" scientific knowledge.[62] This attempt to tame the increasingly obvious structural conflict within Habsburg society created an unprecedented system of strictly enforced food consumption rationalization, which invaded not only workers' private apartments, but also factory owners' villas. As all social structures and relations disintegrated, recourse to the rationalistic discourse of modern science was one of the government's last options to bridge the gap between individual social groups. Faith in scientific authority was one of the last phenomena shared by middle-class scientists, factory owners and their employees alike. This is why the police mercilessly knocked on the doors of workers' apartments, as well as businessmen's villas, where they confiscated the meat they thought was unrightfully consumed.

However, the intensity with which physiology and dietary science entered the public space of the Bohemian lands, and Austria-Hungary as a whole, was also a defensive strategy. It required that the rapid decrease in the food supply be scientifically observed, analyzed and discussed. Trust in a scientific solution for escalating malnutrition, which the experts tried to promote to public, not only reflected the prominent status of science in modern society, but also its fear of failure, which forced its way to the surface with each hunger demonstration or strike. Despite the most modern scientific knowledge on the endless opportunities of food rationalization, the greatest danger to the social order of wartime Habsburg Empire was still a hungry human body.

The Wartime Diet and the Reconstitution of the Working Class

Focusing on wartime society in the Bohemian lands, it is not hard to see which hungry bodies caused the most concern. The structural deficits of the wartime economy drastically intensified material inequalities especially between the city and the countryside. The larger Habsburg towns, often seen as centers of affluence and luxury before the war, changed into places of utter misery and suffering in comparison with the countryside.

One example is the phenomenon of the so-called "rucksack economy" (Rucksackwirtschaft), when especially on weekends the urban population would depart to the areas around large cities with various bags, filling them with groceries during the day and bringing them back in the evening.[63] This occurred all over the Bohemian lands. The social democratic journalist Antonín Nedvěd recalled similar scenes in the surrounding area of wartime Pilsen: "The city's population made unforgettable trips to the countryside for provisions. The scarcer the supplies in the city, the fuller the trains with people leaving for the countryside. At the station on the train line to the agricultural region, usually several hundred people got out and quickly dispersed in various directions. There were men, women and children. All had visible signs of malnutrition and great suffering. … During the potato harvest they dragged remade bags on their backs filled with 50 to 80 kilograms of potatoes. One could only see the bags and, under them, bent over tiny humans like wretched shadows."[64]

In the Bohemian lands, as in the rest of the monarchy, the difference between town and country was visible virtually everywhere. While the Brno University professor František Weyr later recalled how he ate more often at his colleagues' places as the food crisis became worse,[65] Hana Benešová, wife of the future Minister of Foreign Affairs and President Edvard Beneš, wrote in her diary of country life in December 1917, when not even a gram of meat could be found in most Habsburg towns, the following line: "14. XII. [1917] In Mutějovice—pig slaughter—5½ l of milk ½ kg of butter—4 tripe sausages, blood sausage, roughly ½ kg of headcheese and 65 dkg of fresh lard and a cup of cracklings."[66]

The supply crisis brought about an unprecedented rearrangement of social hierarchies—professions reliant on a regular salary were especially susceptible to the rapid decrease of real income that no longer enabled them to satisfy even the most basic needs of their families. Besides the middle and lower bureaucracy, the industrial working class, still mostly working for piece wages, had a prominent position in this new collective of classless people. The government officially acknowledged the food demands of the workers as justly raised by those who carry on their backs a large part of the Austrian wartime production, but at the same time the state system of resources redistribution was not at all able to satisfy these demands.[67] The relatively high social status of the prewar qualified workers was diminished by their decreasing ability to make enough money to feed their families. No longer able to manifest their social status through sufficient consumption, they lost their economic potential. As the prewar hierarchy was overturned, qualified workers often found themselves below the level of the previously even poorer countryside population. Many, however, sank even lower when "there was no choice but to set out for the country and go from farm to farm, homestead to homestead, mill to mill and beg for a bit of flour, eggs, milk or butter and potatoes."[68]

Workers, dependent on salaries that increased much more slowly than prices, were confronted by their descending social status not only during urgent trips to the countryside, but also in everyday life within the limits of their home city and its public spaces. The traditional places where the workers' consumption was manifested, such as markets or pubs, were empty, or the workers played a passive role in them, especially in the second half of the war.[69] One of the major working-class dailies, *Dělnické listy,* wrote about one of the many examples of this on August 1, 1918 in an article:

> A scene in a Bohemian wine-bar. A mother as large as a wardrobe, no—like two wardrobes. In her arms she carries an infant; in her hand a basket. She sits down at a table and orders a quart of wine. "What kind?" "The most expensive kind you have." And then she takes a chicken baked until golden and a slice of bread white as Christmas bread out of her basket. She eats with relish. Jealous stares follow her from all corners of the wine-bar until one of the patrons says: "Mother is having a feast like on the holidays."—"Well, why not?" the mother retorts sharply, "today the farmer is the lord. Before it was the lords in the cities; now they're beggars. Yes, yes, everyone has their time in the sun! Why should I deny anything? We've never had it so good as we do now during the war." And the mother chews her food loudly on purpose. Another patron scathingly remarks: "You keep everything for yourselves and leave us to starve to death, what fine patriots you are!" The mother just curls her lips. "Before the war you called the farmer a buffoon, a hump and who knows what else you called him and now you come after him and ask him to feed you! Well, good for you!"[70]

The urban workers' inability to manifest their social status through appropriate consumption was thus a sign of their radical social descent. Especially in the large industrial metropolises—in the Czech case primarily Prague, Pilsen, or Brno—the wartime economy saw the rapid increase of social inequality, which was felt very painfully in the working-class environment: "Now the peasants are the lords and they make sure that we know it. They have enough bread, cakes, pies, smoked meat, cream cheese, butter, a decent living and so they butter their bread and show it off. Many of them still have some feelings, but many do not give anything and even when they do it is in exchange for tobacco and usually just a small amount. Times have changed, the farmers make profits, they save their money and they eat well. The working class is sinking with malnutrition, has to eat everything that not even a pig would eat and even that they do not have enough of. Pigs today are more valuable than people. They have enough food, nutritious products are given to pigs instead of people, because a nice capital can be made off of them,"[71] the southern Bohemian carpenter Vojtěch Berger noted in his private diary in the fall of 1917.

The complete lack of basic human necessities led to a sharp increase in their prices and the creation of a vibrant black market. Between 1917 and 1918, the price of half a kilogram of flour was often the same as the average daily salary of a qualified worker, and most working-class families had to fight for survival, despite the fact that both parents and all of the older children were employed.[72] At the beginning of the year 1915, the Pilsen social democratic daily *Nová Doba*, for example, repeatedly wrote about the case of the 29-year-old lathe operator Václav CH., who constantly bothered the patrons of local restaurants with his begging. According to the daily, C.H., a qualified worker who had earned a good enough wage before the war to be able to afford his own apartment, had to supplement his income in January 1915 thusly: "He begged the previous evening at house no. 2 in Královská street. When he left, he took the cross that hung in the hallway, which was worth 20K. … Yesterday afternoon he was arrested. In the same house, he had stolen laundry that had been hanging in the courtyard and sold it to fruit-seller Kateřina P."[73]

Packed into Austrian towns, the tens of thousands of human motors that constituted the basic driving force of the wartime economy were at risk of jamming. A large segment of organized workers in the prewar years had learned not to think in terms of the categories of the state ordered nutritional and other norms, but in the categories of seized rights. As a potentially dangerous agent, the working class was therefore the prominent addressee of all the measures issued by the government or employers (these categories often blended together within the Austrian wartime economy) on nutrition and manual labor during all four years of the war, regardless of their preventive or repressive nature. These measures then drastically influenced not only working-class culture, but also the very shape of the working class.

When examining the influence of the wartime politics of food on working-class culture, it is important to be methodologically cautious. The traditional social historiography of the working class sometimes tended to draw clear dividing lines between working-class culture on the one hand, and middle-class or elite culture on the other.[74] The key concepts of the individual discourses, however, often transgressed these seemingly impenetrable lines and changed their meaning depending on the environment. The culture of the lower classes was usually in constant contact with middle-class culture and this exchange helped to shape both cultures.[75] Therefore, to speak *a priori* about working-class culture—an entity in itself that then enters into a relationship with another culture—can be misleading, especially when working-class culture is described in stark contrast to elite culture as popular, traditional, and authentic. The presumed authenticity of the working-class experience is, as in other cultures, only a way to organize the external world into categories that could be easily grasped by the working subjects. These categories, however, may or may not be different from those that are used, consciously or unconsciously, by various elites to make sense of the social world around them.

Therefore, it cannot be said that working-class culture defined itself against the wartime discourse of the rationalization of food. The working class's conception of the relationship between work and food was in many aspects defined within the same framework that determined the scientific horizon, which conceived of the human body as a converter of energy. The dominant metaphor of the human body as a motor that receives fuel and produces work determined the limits of what was thinkable not only in the culture of academic physiology, biology, or dietary science, but also among its objects. Thinking of the working body within the categories of the physical transfer of energy was settled deep within the working class's subjectivity in the second half of the nineteenth century, and remained there during World War I as well. Faith in modern scientific knowledge was shared by both the urban middle-class culture, which was its main producer, and the working-class culture, which was often its object.

At the end of 1914, a thin book appeared among the multitude of working-class literature on the shelves of the "Central Workers' Bookshop" in Prague, titled *Dělnická kuchařka* (A Worker's Cookbook).[76] At first glance, the small volume seemed to be meant for almost every working-class family. The motif on the cover was the gilded price of the book—a mere crown, so that the potential buyer would not actually notice the title of the book at all, but only its unusually low price. However, the book was the second edition of a very successful collection of cooking recipes and aspired to be the basic handbook of every working-class housewife. Its success was helped not only by its low price, but also by the fact that the cookbook provided working-class households with a well-written scientific introduction into the principles of dietary science and rational nutrition.

As the anonymous author wrote, the main goal of the cookbook was to impart the principles of "good cooking." What exactly "good cooking" meant was defined by the cookbook as follows: "The words 'cook well' are to be understood thusly: food must be cooked so as to satiate, but also to provide the body with the amount of material it needs to produce and conserve energy. Chemical analysis of the meals has shown that the meals contain the same material that our bodies are made of. If we look at the body as a cleverly built machine, each small part of which is worn down by work and movement, we realize that food is a fuel that keeps the machine active. Fuel must be good, if the machine is to work well."[77]

After theoretically pondering the purpose of modern nutrition, the cookbook provided specific advice on how to best achieve this "good cooking." It did not immediately mention specific recipes, but provided readers with a very complex list of more than seventy basic foods, from mother's milk to ham, spinach, plums or various types of bread. Each of these items was supplemented by the exact amounts of protein, carbohydrates and fat that the human body could get from them.[78]

Maso obsahuje:

ve 100 gramech	gr. vody	gr. bílkoviny	gr. tuků	uhlohydrátů
Maso hovězí syrové	72·80	20·91	5·19	—
„ „ vařené	74·88	20·—	4·—	—
„ telecí syrové	82·84	15·30	—	—
„ vepřové	68·09	14·—	17·—	—
„ skopové	75·17	14·50	9·—	—
Zvěřina	79·85	18·—	1·—	—
Drůbež	76·22	19·72	1·42	1·27
Krev hovězí	80·82	18·12	0·18	0·03
Šunka uzená	28·11	24·74	36·45	0·16
Sleď	46·23	18·90	16·89	1·57
Vemínko	70·—	28·—	0·40	—

Ostatní potraviny obsahují:

ve 100 gramech	Bílkovin	tuku	uhlohydrátů	vody	soli
Černý chléb	6·11	0·43	46·13	44·58	1·46
Žemle	9·—	1·00	59·50	25·20	1·00
Brambory syrové	1·95	0·15	20·72	73·48	0·95
Hrách syrový	22·85	1·79	52·36	18·97	2·58
Čočka syrová	25·70	1·89	53·46	13·14	3·04
Rýže syrová	7·85	0·88	76·52	13·74	1·01
Krupičná kaše	8·76	2·28	30·—	57·61	1·35
Zelí	1·89	0·20	4·87	90·—	1·23
Květák	2·48	0·34	4·55	90·89	0·83
Mrkev	1·04	0·21	9·40	87·05	0·90
Špenát	2·49	0·58	4·44	89·47	2·09
Med	1·20	0·03	73·72	24·86	0·19
Jablka	0·36	—	12·04	85·60	0·49
Hrušky	0·36	—	11·80	83·33	0·31
Švestky	0·40	—	8·24	86·36	0·66

FIGURE 1.1. Cooking as science: Caloric table for working-class housewives, *Dělnická kuchařka* (Prague, 1914), 9

Everyday shopping at the market became a scientific exercise. Just like chemists and engineers examined the optimal composition of motor fuel in laboratories, the working-class housewife had to carefully consider, while out shopping, how much of which component to add into the meal she would prepare in order to ensure optimal nutrition. Only meals prepared with such insight could lead to the required work output. "We must eat! And if we are properly satiated, we have the strength to carry out our working duties,"[79] laconically stated the famous Czech author of cookbooks for various occasions and consumers Anuše Kejřová in her collection of recipes specifically designed for working-class households. Only a properly satiated worker could properly carry out his or her work, which in turn helped maintain the standard of living of the entire family, intensify the Austrian war effort, and eventually achieve victory over the hated enemy.

At the very end of the war, in May 1918, an in-depth article on the relationship between nutrition and work was published in the main Czech metal workers' periodical, which characteristically summarized the topic of nutrition for a working body:

> A human's life is really a combustion process. … The materials consumed must be replaced with new ones. The body must take in a certain amount carbohydrates, fats and protein so that the consumed amount will be replaced. Even a person who does nothing, does not work and avoids movement needs a certain amount of food, just in order to replace the consumed amount of material. … The amount of nutrition necessary to simply maintain life would be enough, if we were to lie in bed all day. But we work. For this reason, we need more than just the amount of food necessary to keep us alive. Because, just like any other machine, the human organism also is unable to produce new energy, but can only change one form of energy into another. The body transforms chemical energy, which is derived from food, into mechanical energy.[80]

Thus, even within the working-class culture during the war, the human body was seen as a motor that needed a certain amount of fuel for its "standby state," as well as for its basic functions and for work itself, when higher consumption was redeemed by higher work output.[81]

On the other hand, the significance of the specific connection between work and nutrition should not be overstated. Especially at the beginning of the war, when the supply crisis intensified but still had not turned into an acute and universal threat of hunger, working-class discourse was in many aspects still shaped by the traditions of the prewar era. Making demands for better food was generally a substitute topic, through which much more important workers' demands could be articulated. The most important was an increase in salaries, complemented with other claims, such as the shortening of the working day, improving working conditions, hygiene or safety at the workplace.[82]

The issue of decreasing incomes, influenced by a jump in the unemployment rate, dominated the working-class environment at the beginning of the war.[83] The pages of workers' newspapers were filled with more and more cases of factory closings and production reductions during the fall of 1914. For example, the Pilsen daily *Nová Doba* released the results of research conducted by the Czech Association of Unions on the impact of the war on the economy in October 1914. According to these findings, in Bohemia and Moravia since the beginning of the war, 719 factories with 49,051 workers ceased operations. Another 511 factories laid off 18,587 workers.[84] Information from the glass industry revealed 25,000 unemployed glass workers out of a total of 34,000.[85] The light branches of the industry, such as the textile industry, were especially affected. Heavy industry, however, was not spared from the wave of unemployment either. Almost none of the labor unions were able to pay their members' benefits and had to reduce them. The largest Czech federation of metalworkers' unions reduced benefits as early as mid August 1914, when it began to pay benefits only to those members, who had been looking for work for over fourteen days, while travel support was eliminated completely.[86]

The most urgent problem for workers at the beginning of the war was thus not the food shortage, but the decrease in salaries. In fact, in the fall of 1914, few people thought that the conflict would escalate with such intensity and for such a long time, and fears of a supply shortage were very marginal. The major social democratic daily, *Právo Lidu,* for example, assured its readers on August 15, 1914 that "the supplying of our towns was taken care of in a way that instills faith in all of us."[87] Two days later, the editor added another argument: "The level of meat consumption in the empire should not be cause for concern, for we have been producing for export up to now; export is not possible now, however, so there is enough meat."[88] At the same time, the metalworkers' paper assured its readers eloquently that "no matter how long the war lasts, there is no reason to worry about meat. … Hungary, especially southern Hungary, is so rich in pigs and cattle that we can rest assured about this no matter how long the war lasts."[89]

Questions about nutrition became part of the working-class discourse during the first phase of the war only through the dominant prewar demand for raising and maintaining salaries. Their stagnation or decrease in relation to the prices of food, or their complete cessation due to unemployment often resulted in many working-class families not being able to afford as much food as they had been able to before the war, even though there was still enough food in the markets. However, such demands to raise and maintain salaries gradually changed into demands to preserve at least a basic dietary standard. In the eyes of many workers, these demands were targeted at the state and local administrations and particular employers as well.

At the end of August 1914, the Social Democratic Party in Pilsen openly appealed to the city to establish a community kitchen at its own expense, which would provide at least lunches for laid-off workers.[90] In September, these appeals by the workers' political representation began to be heard in Prague,[91] where the local municipal administration immediately began to consider the possibilities of public meal services: "Due to the current economic crisis and the unemployment situation connected to it, the mayor has ordered an inquiry about providing unemployed persons and the poor in Prague with bread and milk by asking the individual large dairies and bakeries if they would be willing to supply Prague with their products and under what conditions."[92]

In Prague during the first two weeks in October, however, the idea of temporarily providing bread and milk was replaced with the more complex concept of feeding the public. The Prague municipal authority quickly picked five places where it would be practical to serve lunches for the city's unemployed and poor population starting in the second half of October.[93] The preparation and consumption of meals was thus in many cases removed from the families and became a matter of public policy. Placing food consumption in the center of public attention opened the door to many situations that reflected the changes in the social order of the Czech, as well as the entire Austrian, society.

When there was still enough food at the beginning of the war, but more and more workers did not have enough financial resources to procure it, the extensive Prague system of public meals was the first sign that the question of food would soon become very important. While the first wave of wartime unemployment receded by the end of 1914, the actual incomes of most of the city's population continued to fall rapidly.[94] In reaction to this urgent social problem, the social practice of eating moved from the sphere of individual households to the sphere of mass consumption in select public places, and the working class first began to regard the issue of food distribution and consumption as a fundamental question that dominated the working-class discourse from the end of 1914 until the end of the war. Workers' newspapers daily printed front-page articles on food supplies, the opening of public kitchens and various other measures. These topics even began to gradually eclipse the reports from the front. The politics of food became one of the central topics around which a new wartime working class restructured itself.

The wartime politics of food virtually disintegrated the prewar working class. First, it moved practically all workers down several rungs in the social ladder. The absence of a decent wage that allowed them to provide for their families destroyed the identity of the collective working class, which was structured around a decent wage for qualified work from which it could derive a standard of material consumption. Most workers were not able to maintain the standard of living of their families, which also included the relative freedom to choose their own food. Second, various nuances in the approach to food and its consumption again reconstituted the working class into several new groups, which were not defined along the lines of income, qualification or age, but according to the ways and things they ate.

At the very bottom was the group that could only eat privately in their own homes, like the Chlupatý and Žižka families who were mentioned at the beginning. While before the war, eating in the comfort of one's own home was a sign of the middle-class social arrangement of separate spheres where the sphere of the home was reserved for reproductive work, including the preparation and consumption of food, during the war eating hidden away in one's kitchen became a sign of utter social decline. This kind of eating at home was inescapably connected with the necessity to procure essential ingredients in public, which was a nightmare for almost all working-class women. Waiting in lines for basic food rations usually took several hours each day, and the food line quickly became one of the primary arenas where a social collective of humiliated individuals that defined itself in opposition to the middle-class order. References to the pervasive injustice in the access to food shaped the language, which allowed criticism of the whole social order based on the daily wait for food rations.

An example of this is the special Christmas edition of *Právo Lidu* from 1916. At Christmastime, when the majority of the population took shelter in their private homes to enjoy at least a semblance of past Christmas feasts, the social democratic paper provoked its readers with a survey that completely overturned the usual Christmas customs. Instead of recipes for the least expensive Christmas fare, the newspaper printed letters from its women readers, which the editors probably intentionally collected in order to publish them all together on the day that was previously synonymous with a period of abundance and satisfaction.

The contrast between the symbolism of Christmas Eve and the realities expressed in the survey was obvious: "It is truly sad and desperate for us mothers. … I received a ration card for half a liter of milk a day for a 2.5-year-old child, valid from the 10th of this month. … Although my child is sick and needs milk, I received half a liter of milk for the whole day. I stood in line from five in the morning until seven. Then I went home; it is impossible to get milk anywhere else. Little Anička wanted to eat, what should I give her for breakfast? Should I give water to this sick child for breakfast, when she has a fever and drinks all day long? What should I give her? Should I lock my sick child inside alone even at night, if I want to get her half a liter of milk? Others have it in abundance without waiting and sick children get nothing? Is this justice?"[95] a working mother signed as "A. F-ová" described her experiences with procuring food within the official system of rations.

The topos of the unjustly suffering, innocent child strongly resonated during Christmastime. It also became a point that connected one's personal feeling of disaffection with the emerging collective of the unjustly suffering and defined this collective against the "others," who unfairly profited from the ration system: "Scene of the action: Prague neighborhood of Žižkov. December, freezing, damp morning, around seven o'clock. A child wrapped in torn up bags is huddling on the ground, holding a cup and waiting for the milkman to open the store. Around him, poor and shabbily dressed women stand shivering in the cold. 'I have to wait it out today. I've been feeding my child black coffee for fourteen days. ... And is this justice? How else should I see it then, when I see that I am unable to get any milk even for my child, and the bank gets a daily delivery of 300 liters of milk for the clerks' snacks and some masters even send their servants out to bring cans of milk home for them.'"[96]

This discourse of injustice naturally also resonated beyond the Christmas holidays. The workers' press increasingly published news about the problem-free food supply of more affluent urban families or farmers in the countryside, and about the continuing discrimination against the poorer working-class families in terms of access to basic foods. The Prague municipal administration's inspections often served as a fine example when they repeatedly uncovered illegal food storages that several banks provided for their employees,[97] extra food rations that the owners of Prague luxury hotels received to support their accommodation businesses,[98] or the affair of the treasurer of the Prague supply office, who allowed certain businessmen to make off with trucks of rationed sugar in exchange for hundreds of thousands in bribes.[99]

The dividing line between those who were constantly waiting in various lines and those from the rest of the urban population deepened. In the summer of 1917, *Právo Lidu* reported on Žižkov milkmen selling milk on purpose only during the hours when most workers would be working the night shift, so that they would not be able to buy milk even though they had enough money and ration cards.[100] In Pilsen, newspapers featured news on how there were not enough cards for workers waiting in bread lines, so their foreheads had to be stamped with indelible ink.[101] The thusly branded waiting workers could not get over the humiliating experience of waiting in line, even afterwards, for they were forced to wear their humiliation visibly for several days everywhere they went.

In the streets of Habsburg towns, a new social hierarchy was created amid the rising supply crisis. At its bottom stood the individuals who had to rely solely on the ration system or their modest self-supply. The currently unemployed workers, or employed workers making minimum wage, mixed with the qualified, better paid workers as well as with the lowest levels of urban poor. However, the shortages in the supply system forced many to depend also on one of the public kitchens that began to crop up at the end of 1914. It was the Prague municipal authority that put in place probably the most complex system of public kitchens in the fall of 1914. In the first half of September, a few town hall clerks traveled to Berlin and Vienna to acquire the basic knowledge of how to establish a system of public kitchens, and in October 1914 the first public kitchen in Prague was opened in the municipal slaughterhouse in Holešovice (nowadays the biggest market hall in the city).[102] The meals in this kitchen were to be handed out only to the unemployed and poor residents of the city, and were funded to a great extent directly by the municipal authority. Only lunches were served, for which the consumers paid merely a symbolic price of 20 halers per meal.[103]

Shortly after opening, interest in these meals exceeded all expectations. In November 1914, the kitchen served over 24,000 lunches every day, and its capacity was instantly exhausted. At the end of 1914 and during 1915, the Prague municipal authority opened three more public kitchens, thereby creating a network of public meals for the poor, which lasted until the end of the war.[104] Although the number of people who took their meals there somewhat decreased at the beginning of 1915, for a large number of Prague workers it continued to be the sole source of warm food. In 1917, the Prague municipal authority still registered over 2.7 million lunches served.[105]

In the beginning, only people who had official proof of unemployment or poverty had access to one of the four Prague kitchens, which became the public site of the working class's social descent. Before the war, the vast majority of the diners was employed in one of Prague's factories and often came from qualified professions that did not allow them to lead an extravagant lifestyle, but still ensured a dignified standard of living.[106] These qualified workers were now subjected to humiliating situations every day. Holding their own fork and knife in their hand, they had to wait in line crammed together with a thousand others, hoping that some lunch would be left when it was finally their turn.[107] All of this under the watchful eye of the police officers and often accompanied by the unpleasant behavior from the kitchen staff.[108]

The actual cooking in the kitchens was influenced by the dominant discourse of maximum food rationalization. Since only the unemployed or socially depressed ate there, it was not necessary to make sure that the food had the right proportion of the basic nutritional components. But even unemployed workers had to receive the established amount of calories. Human motors had to be kept in at least a minimum level of activity in the event that they would be called back to work in the future. The kitchen management was thus set on a "pursuit of calories"; meals were judged solely according to the energy values that they provided for human motors. As Julius Stoklasa and other Czech physiologists and dietary specialists emphasized in their works, one gram of fat, according to the well-known Rubner tables, was able to provide a human body with 9.3 calories, while the same amount of carbohydrates or proteins provided less than half of that amount. The public kitchens therefore saw fat as the most effective source of calories. At the beginning of October 1914, the Prague municipality therefore approved a proposal to collect drossy suet from Prague military kitchens, which was to be used to prepare the meals.[109] Almost immediately there were complaints about the inedible lunches. Among the first to complain were the workers Rudolf Smolík and J. Popelka, who pointed out that cooked suet, served without a piece of bread, is practically inedible. However, their complaint, like all the others, was rejected with the explanation that the lunches served closely complied with the requirements for adequate nutrition, and there was therefore no reason to change anything.[110]

Although the portions of drossy suet complied with the prescribed caloric table values, the complaints continued unabated. The Prague municipal authority therefore had to implement more drastic measures. In late October 1914, anyone who complained about the quality of the lunches served was banned from entering the kitchens within the city's network.[111] This ban temporarily silenced the negative opinions, but did not quite deal with the actual situation in the kitchens. The municipal committee did not receive any official complaints for a while, but there were more and more incidents in which policemen, or even a doctor, had to intervene in the kitchens.[112] The situation became untenable during the summer and fall of 1916, when the consequences of the supply crises began to manifest themselves with ever greater intensity, and more and more people who had some money, but nothing that they could buy with it, were forced to eat in the kitchens.

The original concept of public kitchens as places where only the poor and unemployed could eat began to fall apart, and the public kitchens changed into places where a wider segment of consumers had to be provisioned for. The new social composition of the consumers came with a new set of problems. At a meeting of the Prague committee for public consumption on October 20, 1916, council elder Ferdinand Kellner stated: "Complaints continue to come in about the bad and disgusting lunches, which often stank."[113] The situation in the kitchen in the Prague neighborhood of Holešovice became so dire that there was a proposal to shut it down.[114] The proposal was decisively rejected, and the relevant municipal committee further persisted in its position that the kitchen was fulfilling its purpose in providing the unemployed and poor Prague workers with necessary calories. According to the committee, the food itself was not of the best possible quality, but it nevertheless provided the energy needed just as well as more expensive food.

According to the Prague municipality, the only problem that needed to be dealt with in terms of the quality of the food was not the quality of the meals themselves, but the hygienic conditions in the kitchens. The largest public kitchen in the central slaughterhouse of Prague was again the most problematic. The storage of old hooves, dried blood and a herd of 27 live pigs, which the manager of the kitchen, Syrovátka, was keeping directly in the dining rooms, threatened the hygienic standard of the cooking, according to the responsible officials. Nevertheless, once these most blatant problems were addressed, the kitchen continued working until the end of the war.[115]

The everyday experience of humiliation in order to prove the right to repugnant food, served under the watchful eye of the police and in unappealing hygienic conditions, publicly represented the new position of a segment of the industrial working class in Czech society during the war. Allowing not only the unemployed, but also many of the employed workers, to use the kitchen, which occurred in 1916, blurred the dividing line between employed/unemployed, and replaced it with a line between those who did not have the option to eat anywhere else but in the public kitchens or at home, and those who ate elsewhere. Such a situation became possible in Prague in the summer of 1916, when the Prague municipal authority opened a second network of public kitchens, this time officially described as kitchens for the "middle classes."[116]

The idea to open a second network of public kitchens not for unemployed and poor workers, but for the middle class, floated among municipal officials already in the first half of 1916, when more and more Prague citizens were affected by the supply crisis. The initial idea was to expand the already existing network of public kitchens so that they could accommodate more eaters. This proposal, however, was met with the opposition of several Prague council members, who were afraid of the humiliation of consumers, who this time would not be just poor workers. Council member Václav Brož aptly formulated this fear at a session of the Prague supply committee. According to him, the "mass eating of these people within the public kitchens would not be appropriate. Public kitchens are effective, but they are after all considered to be organizations for the poor, and their enlargement would not be the right thing. … This is about the other persons, who are too embarrassed to enter the public kitchens. This is a large group of people, for whom the four current kitchens are not enough."[117]

The newly established middle-class kitchens were not supposed to provide subsidized meals, but meals at a price that would cover the costs of production. The prices were to be reduced by the centralized acquisition of the ingredients, which the town hall wanted to achieve with bulk discounts that would guarantee lower prices. Unlike the kitchens for the poor, the middle-class kitchens were to be open to anyone without having to fulfill special requirements, and the Prague municipal authority insisted that "the collective dining of the middle classes should not be merged with the people's kitchens. These should remain kitchens for the poor, and a special kitchen should be established for the middle classes."[118]

These plans were fulfilled in the second half of 1916, when five of these middle-class public kitchens were opened, with locations all over Prague. The Old Town was served by the U Vejvodů restaurant in Jilská street; New Town by the U Palmů restaurant on Karlovo náměstí (Charles Square) in house number 319/3 and the Měšťanská beseda (today Vladislavova street 20). The densely populated, mostly working-class areas outside of the historic center of the city were served by the kitchen in Bělského třída in Holešovice (today Dukelských hrdinů street) and the U Deutschů restaurant in Primátorská třída in Libeň (today the building of the Pod Palmovkou theater in Zenklova street).[119]

The middle-class kitchens differed from the people's kitchens not only in terms of the price of the meals offered. All of the serving places had to adhere to the "restaurant guidelines" that were posted in every kitchen on a visible spot. Unlike the people's kitchens, the staff was strictly supervised and had to behave "correctly and decently to all clients."[120] Lunches were served based on advance payment, so the number of eaters was known a day in advance, which prevented long lines and uncomfortable jams. For this reason, there were also no police present in the serving places. For each of the five kitchens there was one inspector, who was constantly present and would continuously check everything that was happening in the kitchen.[121]

While the people's kitchens used mostly waste or the cheapest ingredients, the middle-class menu was noticeably more attractive. In 1916, diners in the middle-class kitchen ate beef twenty-seven times, dumplings with cabbage seven times and once even rabbit.[122] In 1917, when meat practically disappeared from the counters of regular shops, the municipal committee managed to obtain regular deliveries of meat for these kitchens from the Prague slaughterhouses. In February 1917, for example, the Prague municipal authority decided to allocate a significant amount of beef stomachs and heads to middle-class kitchens, while the people's kitchens had to make do with only several bags of peas.[123] In the last years of the war, the middle-class kitchens were actually the only place where it was possible to get meals of meat.[124]

If we look at the geographical distribution of the kitchens, it is apparent that the Prague municipal authority from the beginning assumed that a significant segment of Prague workers would belong to the "middle class" that would take their meals there. The Libeň and Holešovice kitchens were surrounded not only by many factories, but also by large working-class housings. Also, the "restaurant guidelines" of these kitchens, which explicitly prohibited the serving, consumption, or any other procurement of alcoholic beverages, reveal to what extent the stereotypes of the alcohol-obsessed working class were ingrained in the opinions on the presumed clients.[125]

Immediately after they opened, the middle-class kitchens indeed became places where collectives that would have been inconceivable before the war were formed. In 1916, the kitchen in the U Vejvodů restaurant, for example, hosted more than 400 male and female workers, as well as over 380 clerks, some Prague policemen, self-employed tradesmen, and even several Prague lawyers, journalists, and academic sculptors.[126] The Prague municipality was aware from the very beginning that opening this second network of public kitchens would result in the redrawing of the prewar social dividing lines. In all official statements, the town hall referred to them as kitchens for the "middle class." However, after the war, Prague officials admitted without scruples that "these were kitchens for a wide segment of the society, which were not supposed to have the character of an establishment for the poor, because everyone should have received a lunch at overhead prices. Yet these kitchens had a considerable charitable and social nature."[127] The purposeful labeling of the kitchens as places for the "middle class" and their arrangement, which respected middle-class values of dignified dining within the tidy environment of established Prague restaurants—no waiting, in an orderly fashion and without the excessive consumption of alcohol—erased this initial charitable purpose of the kitchens, and, as a result, created an entirely new "middle class" that had to be distinguished from the most depressed segments of the urban poor.

In locations where there was a risk that diners in the middle-class kitchens would commingle with consumers from the people's kitchens, the Prague town hall made sure that both types would be as far away from one another as possible, so that the clients of the kitchens could not be confused. This problem occurred primarily in neighborhoods, where there was a large concentration of workers. While before the war these neighborhoods were, in the imagination of most middle-class politicians, inhabited by a monolithic mass of workers, during the war this mass was divided into several new groups, one of which became a significant part of the new "middle class" for which new kitchens were opened. Thus, for example, in the Prague neighborhood of Holešovice where there were both types of dining rooms, the middle-class room was located in Bělského třída—

more than one and a half kilometers away from the people's kitchen in the Prague slaughterhouse.

From the perspective of the Prague municipal authority, the opening and running of the second network of kitchens in 1916 meant the continuing transcription of the new wartime social organization into an understandable spatial and institutional framework. The system of these public kitchens created an entirely new "middle class" and provided a previously non-existent opportunity for the creation of hitherto impossible social networks that dismantled many of the prewar social barriers. The urban society lost its prewar divisions, usually established along the lines of income and profession, replacing them with a new stratification, in which individuals were categorized according to what and where they ate. In the middle-class kitchens, qualified workers mixed with the white-collar office workers, as well as with the traditional middle-class professions, such as lawyers and academics. They were therefore in a significantly different situation from their less fortunate colleagues, who were dependent on begging or on consuming scraps from one of the people's kitchens.

At a time when the war totally confused the income structure of the population, the physical act of eating in a public arena was one of the few ways in which it was possible to determine someone's social position. The Prague network of public kitchens was one of the most complex systems in the Bohemian lands, though not the only one. Since the beginning of 1917, many delegations from all the corners of Austria had been coming to Prague, collecting information as well as practical hints, and subsequently implemented the Prague model in their communities.[128]

However, there was still one segment of the working class that did not use any of the public kitchens and thus remained outside of these newly emerging networks. This rather large group of workers was employed in factories that partly before the war, but mainly during the war, implemented their own meal programs. There were many similar businesses in the Bohemian lands. Already in the fall of 1914, business inspectors reported of their inspection trips all over Bohemia that "in many enterprises meal plans are put in place, thanks to which workers are provided with hearty meals for a small fee or entirely for free. One sugar refinery, thanks to the company's significant support, provided three meals a day for 46 halers, consisting of coffee for breakfast and dinner, and soup and a large portion of veal and dumplings for lunch."[129]

As a general rule, various forms of company meal plans were provided especially by larger businesses that were essential to the wartime economy or had good contacts within the political administration. Many of these companies, which were either directly militarized or placed under the law on state operated enterprises, were located in the Bohemian lands.[130] Militarized enterprises had direct access to the central supply system, and some of the other businesses also managed to lobby the local administrations and military offices for a preferential supply of food, which they then distributed among their employees. The Austrian business inspection could thus state, for example, that during its inspections in 1916 in the third Prague inspection district (the area of Karlín, Český Brod, and the eastern part of Central Bohemia), almost all of the local heavy machinery factories had their own canteen. [131] In the Pilsen area, the Škoda factory led the pack, having provided its workers with food since the fall of 1914, while the largest Pilsen brewery joined it at the end of the year.[132] In the Pardubice area, employee canteens were opened even in enterprises that were not directly involved in the Austrian war production, such as one local laundry, which opened its own canteen for roughly 300 workers already at the end of 1914.[133] Employee canteens were not just places where hundreds of industrial workers were satiated, but also places where they could be easily disciplined. Each worker was subject to the same meal regulations—they all ate at the same time and consumed the same food, which they had to finish eating by a strictly determined amount of time.[134] Factory canteens were also a weapon against workers' protests, for the first thing that the factories did when strikes or other protests broke out was to close the factory canteens, which were the only sources of food for many workers during the weekdays.[135]

Workers employed in factories with meal programs did not go to the public kitchens, and their family members often did not have to stand in long food lines. Many factories not only calculated employee food rations for individual workers but were also willing to hand out certain foods for the appropriate number of ration cards to their family members as well. In fact, many enterprises behaved similarly to municipal administrations, providing their employees with meals or foods at significantly subsidized prices.

As in the cities, establishing canteens in industrial production sites clearly reflected an inner social order. The quality of the meals, the location of the canteens and their equipment reflected the internal business hierarchy, similarly to variously differentiated networks of kitchens in the public places of larger towns. The largest factory in the Bohemian lands—Pilsen's Škoda armament factory—introduced the most distinctive demonstration of this internal corporate hierarchy.

The biggest enterprise in all of Bohemia separated manual laborers from the administration personnel during mealtime.[136] The workers thus ate separately from accountants, cashiers, inspectors, and other administrative professions, whose meals were usually prepared with greater care. Even the spaces in which they took their meals were better maintained than the regular workers' canteens, where, according to eyewitnesses, "there is disorder and filth. Despite the fact that there is a large enough kitchen staff, the right care is not paid to the preparation of meals, so that the rank and file often receive disgusting, unfit for consumption swill for their money. Leftover food is set aside, lies on the ground or in dirty receptacles, and is thrown back into pots and resold. Fish, from which any cook could prepare a delicious meal, were sold to the workers in a disgusting, mushy porridge. ... Bread lies on open counters in the canteen among the steam, smoke and dirt; it often lies on the ground, so the worker receives it all sodden, sour and soiled. It has also often been nibbled on by mice."[137]

Workers in enterprises with their own canteens on the one hand did not have to worry so much about where to get food. Nevertheless, these workers were generally subjected to pervasive symbolic representations of the new social order, just like their colleagues, who dined in one of the public kitchens in the city. Those with respectable jobs before the war, used to a decent salary, were publicly degraded to the lowest place in the internal hierarchy of the enterprise. As human motors, they were eligible for certain supply programs, but due to the growing food shortage these programs were only aimed at satisfying their demands for energy. The distribution of food through employee canteens, just like communal kitchens for the poor, was limited to the mechanical application of nutritional guidelines, which prescribed precisely established amounts of calories for manual labor. Any other nutritional context was forced out by the dominant discourse, which saw the human diet only as a part of the physical energy transfer.

In September 1916, the Vienna Ministry of War sent out an inquiry to all Czech enterprises under military administration about the state of their supplies and requested that they report their food requirements for the following year.[138] More than a month after this mass inquiry was sent out, however, only a few responses made their way back to Vienna. Most of the military administrators had to be vehemently encouraged and notified that the ministry's questions were not rhetorical, and that clear responses are expected of each of them.[139] Only after this second strong appeal did the administrators of the Czech militarized businesses react in greater number. Practically all of them reported the total absence of meat, which gradually disappeared from the working-class diet, and the rapid rise in the prices of other food, which made it harder for many workers to get the variety of foods that they were used to before the war.[140] However, this simplification of the diet was not perceived as a barrier that would prevent workers from carrying out their work obligations. According to the opinion of the military administrators, the vast majority of workers had a guaranteed supply of food, which delivered all the necessary energy for their work to their bodies. Therefore, most of the military administrators did not consider it necessary to react to the first appeal by the ministry in Vienna. For instance, the military administration in Most (Brüx), one of the central brown coal regions in Austria, reported that not only the meat supply was at a standstill, but the supply of fat as well. However, potatoes, cabbage, beets, and bread provided the necessary nutrition.[141] A similar report arrived from the northern Bohemian town of Trutnov, where the military administrator reported the shortage of legumes and fats, which, in his opinion, did not have an impact on wartime production, because the energy needs of the local workers were covered by other foods, primarily bread and potatoes.[142]

The situation changed radically in the first months of 1917, when in many factories, due to the bad harvest, not only the supply of meat and fats, but also of basically everything, was limited or totally stopped.[143] In the Most coal district, the supplies of potatoes and bread began to dwindle in February, while in some of the smaller mines there was no centrally distributed food at all.[144] Roughly 30,000 workers and miners from the north Bohemian coal districts of Loket, Karlovy Vary, Sokolov, Chomutov, Teplice, Ústí nad Labem, Bílina, and Most did not have any more potatoes by mid February 1917, and the number of cases of diarrhea and general physical collapse increased.[145] According to the Most regional authorities, mining was in immediate danger of ceasing in the entire northern Bohemian coal area in March: "I have recently repeatedly warned that unless immediate assistance will be provided, coal mining will completely cease within 14 days," reported the responsible military person on the spot to the headquarters.[146]

It was at this moment, in February 1917, that the food supply began to collapse, and the central military administration began to deal with the situation. As can be seen from the rapid response, a solution still existed. Already twelve days after the desperate appeal from the Most district, the worst holes in the supply chain were patched up and the local workers again had at least minimal sources of energy and were able to continue their work. Supplies of potatoes and flour were immediately renewed, and the central administration in Vienna looked for ways in which to better ensure the supply to the northern Bohemian coal district, which was hitherto relegated to supplies solely from the Bohemian lands.[147] The government in Vienna thus recognized the utmost importance of this area for the Austrian war effort, and at least for the first half of 1917 ensured a sufficient food supply, which helped the local, extremely valuable human motors to keep working. Although sources of protein continued to be absent, the extra delivery of hundreds of truckloads of potatoes was meant to create an emergency supply of nutrition, which provided at least some of the energy necessary for further work.[148]

However, not even from the perspective of the respective military administration was this a definitive solution; it was a state of emergency, in which human machines could not work for very long. The limited food supply, which was devoid of meat and many other basic foods, meant that the necessary amount of energy had to be attained by flour-based products, sugar, and potatoes. Their limited amount, coupled with the absence of other food, was not able to generate enough energy, so it was necessary to further count on possible work failures: "As far as the legal prescriptions are concerned, miners are receiving only the rations of bread flour and sugar. They do not have access to any other food. Being manual laborers, they are unable to work with such a supply for very long,"[149] concluded the Teplice local authorities in May 1917, and pointed out that under such supply conditions, it is only possible to provide emergency operations in all of the local factories.

Industrial production was switched to a similar "state of emergency" during 1917 not only in northern Bohemia but also in the entire Bohemian lands. In March 1917, youths aged 15 and 16 years had to be summoned to the industrial factories in České Budějovice (Budweis) and its surroundings, because neither physical strength nor the number of older colleagues was enough to fulfill the production needs any longer. From the perspective of the Austrian economy, this was very uncomfortable, for according to the available knowledge it was necessary to provide adolescent men in particular with more energy than the adults. When it was not possible to get anything even with food ration cards, the pervasive lack of basic foods led to many unrests, when these youngsters, who had been quickly sent to work in local factories, refused to content themselves with the insufficient supplies they received at work.[150]

The larger industrial factories often had a more prominent status in the supply system. However, when food began to grow scarce in the entire monarchy, not even the military commanders or the civilian managers of these factories had any special channels to get at least the basic foods for their employees. A more complicated way of providing food for the intended factory consumers often meant that a portion of it would get lost along the way from central storerooms to the factory management and the persons responsible for the factory canteens. The workers of these enterprises were spared the humiliating experience of procuring basic foods in the public space but, as a result, often received a smaller supply of food that got lost in the complicated chains of distribution.[151] The growing supply crisis, which reached its peak in the last two years of the war, also affected the workers, who were partly or completely dependent on their employers for their food supply. Even here, the collapse of the Austrian central system affected the very core of the prewar workers' movement, when the vast majority of the workers found themselves at the bottom of the new social order of the home front.

The wartime politics of food thus effected a deep transformation of the working class. Many studies devoted to the social transformation of Czech society during the second half of the long nineteenth century emphasize the creation of collective identities in the sphere of production. According to those studies, the working class represented a group of individuals connected by a common position within the process of economic production, which, with the help of various mechanisms and strategies, began to transform itself into a conscious, politically active "working class" at the end of the nineteenth and the beginning of the twentieth century. Thus, according to this interpretation, the working class became a collective historical agent, which acted in the defense of its own economic interests.[152]

Such an approach often overlooks the varied interests of individuals who were not only defined by their position in the economic structure of the state but also by the salary they received from it. Recent studies have shown that different salaries, or work qualifications, did not necessarily have a major significance in terms of shaping the social positions of the laboring men and women. The social stratification of industrial societies, or "class membership," is something that people did not automatically take away from the factory gates or the office doors—it required the influence of the discourses representing structural differences. Different salaries, standards of living, and even social barriers often became valid only once they were "experienced" in the form of public consumption, which did not always reflect structural socioeconomic differences.[153]

In the wartime society of the Habsburg monarchy, however, the most important commodity was food. It was precisely the consumption of food that conditioned the emergence of a new social hierarchy, and, with it, deeply transformed the working class. Yet the consumption of food did not exist in a vacuum. A coherent discourse of modern science determined what food was, and why and how it should be eaten, and it viewed the human body, already before World War I, as a type of motor—a mechanical converter of physical energy into manual labor. This metaphor dominated thinking about the human body and its relationship to labor, enabling the emergence and operation of a "politics of food" that completely transformed the working class during the four years of the war.

If we understand class as the location of individuals within a social hierarchy, defined by material wealth, non-material privileges, and the amount of access they have to various resources, then the wartime politics of food made the prewar working class practically disappear. People who stood for hours in lines for bread, butter, milk, or potatoes were often employed in the same factories as many of the people who ate at the public middle-class kitchens, even if they belonged to completely different social groups. The workers who remained behind the walls of the factories even during lunch breaks and consumed pre-cooked meals in the factory canteen made up a distinct group. The collapse of the prewar social order meant the collapse of an organized working class. The old working class was split up and reconfigured within the new wartime hierarchies, shaped by the rationalizing discourse of modern science. These hierarchies were created not only in factory production halls, but also while workers waited endlessly in food lines, or while sitting on long benches on which hundreds of wretched bodies were packed together to eat wasted fats, or while dining in middle-class restaurants where rabbit could be served on white tablecloths.

Notes

1. Karl Kraus, *Die letzten Tage der Menschheit,* Act 3, Scene 10:
   Wenn ich mir was wünschen sollt,
   Ich wüsst schon lange, was ich wollt,
   Ein Knödel müsst es sein,
   Aus Semmeln, gut und fein!
2. Archives of the Capital City of Prague (henceforth ACCP), Referát XIV. Zásobování, karton 99, sign. e10/601-900, Jelínek.
3. The "meatless days" were introduced in the whole of Austrian Cisleithania already during May 1915. Selling of meat and all meat products was forbidden on two days a week, usually on Tuesdays and Fridays. *Reichsgesetzblatt für die im Reichsrate vertretenen Königreiche und Länder* (henceforth RGBl) (May 9, 1915): 189. By the decrees from June and September 1916, this was extended to all Mondays, Wednesdays, and Fridays. RGBl 1916, 513–15 and 755.
4. RGBl, 515, § 13.
5. ACCP, Referát XIV. Zásobování, karton 99, sign. e10/601-900, Jelínek.
6. Ibid.
7. ACCP, Referát XIV. Zásobování, karton 99, sign. e10/601-900, Chlupatá and Žižková.
8. Ibid., Chlupatá.
9. Ibid., Žižková.
10. On the wartime Austro-Hungarian supply system, see Hans Loewenfeld-Russ, *Die Regelung der Volksernährung im Kriege* (Vienna and New Haven, 1926); Juraj Křížek, "Die Kriegswirtschaft und das Ende der Monarchie," in *Die Auflösung des Habsburgerreiches. Zusammenbruch und Neuorientierung im Donauraum,* eds. Richard Georg Plaschka and Karl-Heinz Mack (Vienna, 1970), 43–52. On German inspirations, see Friedrich Zunkel, *Industrie und Staatssozialismus. Der*

Kampf um die Wirtschaftsordnung in Deutschland 1914–1918 (Düsseldorf, 1974); Hans Gotthard Ehlert, *Die wirtschaftlichen Zentral-behörden des Deutschen Reiches 1914 bis 1919. Das Problem der "Gemeinwirtschaf" in Krieg und Frieden* (Wiesbaden, 1982).

11. Maureen Healy uses this term, but all studies confirm in various ways the central role that the food shortage played in the Austrian lands. Arthur J. May, *The Pasing of the Hapsburg Monarchy 1914–1918*, vol. 2 (Philadelphia, 1966); Reinhard J. Sieder, "Behind the Lines: Working-Class Family Life in Wartime Vienna," in *The Upheaval of War: Family, Work and Welfare in Europe, 1914–1918*, eds. Richard Wall and Jay Winter (Cambridge, 1988), 109–38. For special emphasis on the Bohemian lands, see Peter Heumos, "Kartoffeln her oder es gibt eine Revolution. Hungerkrawalle, Streiks und Massenproteste in den böhmischen Ländern 1914–1918," *Slezský sborník* 97, no. 2 (1999): 81–104; Ivan Šedivý, *Češi, české země a velká válka 1914–1918* (Prague, 2001); Jan Havránek, "Politické represe a zásobovací potíže v českých zemích v letech 1914–1918," in *První světová válka a vztahy mezi Čechy, Slováky a Němci*, eds. Hans Mommsen, Dušan Kováč, and Jiří Malíř (Brno, 2000), 37–52.
12. Second Letter of Paul to the Thesalonikians 3, 10. See Jakob Tanner, *Fabrikmahlzeit, Ernährungswissenschaft, Industriearbeit und Volksernährung in der Schweiz 1890–1950* (Zürich, 1999), 11–19.
13. Fabio Bevilacqua, "Helmholtz´s Über die Erhaltung der Kraft: The Emergence of a Theoretical Physicist," in *Hermann von Helmholtz and the Foundations of Nineteenth-Century Science*, ed. David Cahan (Berkeley, 1994), 291–333; Ingo Müller, *A History of Thermodynamics: The Doctrine of Energy and Entropy* (Berlin, 2007), 47–78.
14. Donald S. L. Cardwell and James Joule, *A Biography* (Manchester and New York, 1989), 215–28.
15. Anson Rabinbach, "Ermüdung, Energie und der menschliche Motor," in *Physiologie und industrielle Gesellschaft. Studien zur Verwissenschaftlichung des Körpers im 19. und 20. Jahrhundert*, eds. Philipp Sarasin and Jakob Tanner (Frankfurt am Main, 1998), 293–94.
16. Anson Rabinbach, "The European Science of Work: The Economy of the Body at the End of the Ninteenth Century," in *Work in France: Representations, Meanings, Organization, and Practice*, eds. Steven L. Kaplan and Cynthia J. Koepp (Ithaca and London, 1984), 475–513.
17. Carl Ludwig, "Leid und Freude der Naturforschung," in *Carl Ludwig. Begründer der messenden Experimentalphysiologie 1816–1895*, ed. Heinz Schröer (Stuttgart, 1967), quoted in Maria Osietzki, "Körpermaschinen und Dampfmaschinen. Vom Wandel der Physiologie und des Körpers unter dem Einfluß von Industrialisierung und Thermodynamik," in *Physiologie und industrielle Gesellschaft. Studien zur Verwissenschaftlichung des Körpers im 19. und 20. Jahrhundert*, eds. Philipp Sarasin and Jakob Tanner (Frankfurt am Main, 1998), 313.
18. See Bruno Latour, *We Have Never Been Modern* (Cambridge, MA, 1993), 62–65.
19. Jules Amar, *Le moteur humain et les bases scientifiques du travail professionnel* (Paris, 1914). For the sake of this book the author made use of the English ver-

sion: Jules Amar, *The Human Motor, or The Scientific Foundations of Labour and Industry* (London, 1927).
20. Anson Rabinbach, *The Human Motor: Energy, Fatigue and the Origins of Modernity* (Berkeley and Los Angeles, 1992), 2–5.
21. Tanner, *Fabrikmahlzeit,* 30.
22. Ibid., 15–19.
23. Richard Whitley, "Knowledge Producers and Knowledge Acquirers. Popularization as a Relation Between Scientific Fields And Their Publics," in *Expository Science: Forms and Functions of Popularisation, Sociology and Science,* eds. Terry Shinn and Richard Whitley, vol. 9 (Dordrecht, Boston, and Lancaster, 1985), 3–28.
24. Philipp Sarasin, *Reizbare Maschinen, Eine Geschichte des Körpers 1750–1914* (Frankfurt am Main, 2001), 146.
25. RGBl 1914, 1113.
26. Ibid., 1165–66.
27. RGBl 1915, 84–85.
28. Ibid., 130.
29. Pavel Scheufler, "Zásobování potravinami v Praze v letech 1. světové války," *Etnografie dělnictva* 9 (1977): 159–63.
30. Šedivý, *Češi, české země a velká válka,* 244.
31. Maureen Healy, *Vienna and the Fall of the Habsburg Empire. Total War and Everyday Life in World War I* (Cambridge and New York, 2004), 33.
32. On the importance of the concept of "Durchhalten" for cementing the social consensus of the wartime societies, see Arnd Bauerkämper and Elise Julien, eds., *Durchhalten! Krieg und Gesellschaft im Vergleich 1914–1918* (Göttingen, 2010).
33. Emil Abderhalden, *Die Grundlagen unserer Ernährung unter besonderer Berücksichtigung der Jetztzeit* (Berlin, 1917), 2.
34. Antonín Merhaut, *Zásobování a výživa lidu v době válečné* (Prague, 1916), 4.
35. František Mareš, *Výživa člověka ve světle fysiologie* (Prague, 1916), 36.
36. Max Rubner, *Die Gesetze des Energieverbrauchs bei der Ernährung* (Leipzig, 1902).
37. Julius Stoklasa, *Výživa obyvatelstva ve válce!* (Prague, 1916), 10.
38. Abderhalden, *Die Grundlagen unserer Ernährung,* 98–99.
39. Carl von Voit, *Physiologie des allgemeinen Stoffwechsels und der Ernährung* (Leipzig, 1881), 518–28.
40. Armand Gautier, *Diet and Dietetics* (London, 1906), 91–98.
41. http://en.wikipedia.org/wiki/List_of_countries_by_food_energy_intake – Reference no. 4: http://www.fao.org/fileadmin/templates/ess/documents/food_security_statistics/FoodConsumptionNutrients_en.xls. Accessed January 9, 2013.
42. Voit, *Physiologie des allgemeinen Stoffwechsels,* 525.
43. Rabinbach, *The Human Motor,* 128–33.
44. Loewenfeld-Russ, *Die Regelung der Volksernährung im Kriege,* 335; Hans Hautmann, "Hunger ist ein schlechter Koch. Die Ernährungslage der österreichischen Arbiter im Ersten Weltkrieg," in *Bewegung und Klasse. Studien zur österreichischen Arbeitergeschichte,* ed. Gerhard Botz (Vienna, 1978), 661–82.
45. Šedivý, *Češi, české země a velká válka,* 252.

46. Osietzki, "Körpermaschinen und Dampfmaschinen," 341–46.
47. Cf. Massimo Montanari, *Hlad a hojnost. Dějiny stravování v Evropě* (Prague, 2003), 152–55; Tanner, *Fabrikmahlzeit*, 71–78.
48. Merhaut, *Zásobování a výživa lidu*, 8.
49. Julius Stoklasa, *Das Brot der Zukunft* (Jena, 1917), 113.
50. Ibid., 101.
51. Mareš, *Výživa člověka ve světle fysiologie*, 83.
52. Josef Kafka, *Úsporná výživa. Hospodářská a zdravotní reforma výživy, jídelního lístku, vaření a zažívání* (Prague, 1915), 41.
53. Mareš, *Výživa člověka ve světle fysiologie*, 45–46.
54. On the scientification of cookery during and before World War I, see Caroline Lieffers, "'The Present Time is Eminently Scientific': The Science of Cookery in Nineteenth-Century Britain," *Journal of Social History* 45, no. 4 (2012): 936–59; Viera Hierholzer, *Nahrung nach Norm. Regulierung von Nahrungsmittelqualität in der Industrialisierung 1871–1914* (Göttingen, 2010), 33–47.
55. Kafka, *Úsporná výživa*, 59.
56. Merhaut, *Zásobování a výživa lidu*, 4–5.
57. Stoklasa, *Das Brot der Zukunft*, 174.
58. Mareš, *Výživa člověka ve světle fysiologie*, 47.
59. Stoklasa, *Výživa obyvatelstva ve válce!*, 18.
60. On the contemporary debatte about surrogate food, see Martin Franc, "Výmysly německých profesorů i návrat k zkušenostem předků. Přírodní potravinové náhražky za 1. světové války v českých zemích," *Práce z dějin Akademie věd* 4, no. 1 (2012): 1–14.
61. Tanner, *Fabrikmahlzeit*, 53–60.
62. Lukáš Fasora, *Dělník a měšťan. Vývoj jejich vzájemných vztahů na příkladu šesti moravských měst 1870–1914* (Brno, 2010), especially 97–103.
63. Healy, *Vienna and the Fall of the Habsburg Empire*, 43–61; Sieder, "Behind the Lines," 112–13.
64. Antonín Nedvěd, *Jak to bylo na českém západě 1914–1918. Záznamy a dokumenty* (Pilsen, 1939), 120.
65. František Weyr, *Paměti. I. Za Rakouska (1879–1918)* (Brno, 1999), 434.
66. The National Museum Archives, Deník Hany Benešové, fond Edvard Beneš, karton 2, zápis ze 14. prosince 1917, in *Válečné deníky Edvarda a Hany Benešových (1915–1918)*, ed. Dagmar Hájková and Eva Kalivodová (Prague, 2012), 27.
67. On the worsening supply situation of Austria-Hungary that heavily contributed to its eventuall collapse see Loewenfeld-Russ, *Die Regelung der Volksernährung im Kriege*, 52–71; Healy, *Vienna and the Fall of the Habsburg Empire*, 31–86; Šedivý, *Češi, české země a velká válka*, 250–259; Richard Georg Plaschka, Horst Hasselsteiner, and Arnold Suppan, *Innere Front. Militärassistenz, Widerstand und Umsturz in der Donaumonarchie 1918* (München, 1974), vol. 1, 53–61.
68. Nedvěd, *Jak to bylo na českém západě 1914–1918*, 131.
69. See for example *Večerník Práva Lidu* 285 (December 16, 1916): 4.
70. *Dělnické listy* 1. 8. 1918. Cited from ACCP, Paměti Vojtěcha Bergera, Z Války, Kniha II., politická, od 15. ledna 1914 do 4. října 1919, 202–03.

71. ACCP, Paměti Vojtěcha Bergera, Z Války, Kniha III., od 12. února 1917 do 26. ledna 1918, zápis z 15. října 1917, 138.
72. Cf. Vladimír Průcha, "Nástin vývoje nominální mzdy zaměstnaného průmyslového dělníka v Československu v letech 1913–1917," *Sborník historický* 13 (1965): 65–91.
73. *Nová Doba* (January 15, 1915): 3.
74. For the Bohemian context see Jana Machačová and Jiří Matějček, *Nástin sociálního vývoje českých zemí 1781–1914* (Prague, 2010), 145–377.
75. Fasora, *Dělník a měšťan*, 261–347.
76. *Dělnická kuchařka* (Prague, 1914).
77. Ibid., 5.
78. Ibid., 9-10.
79. Anuše Kejřová, *Dělnická kuchařka se zřetelem na malé dělnické domácnosti* (Hradec Králové, 1914), 5.
80. The article "Výživa a práce" (Nutrition and Work), *Kovodělník* 18 (August 23, 1918): 72–73.
81. See the article "Únava při práci" (Fatigue during Work), *Kovodělník* 37 (September 16, 1915): 182.
82. Michael Mesch, *Arbeiterexistenz in der Spätgründerzeit—Gewerkschaften und Lohnentwicklung in Österreich 1890–1914* (Vienna, 1984), 84–120; Fasora, *Dělník a měšťan*, 39–77.
83. Cf. Šedivý, *Češi, české země a velká válka*, 222–23.
84. *Nová Doba* (October 9, 1914): 3.
85. *Nová Doba* (October 20, 1914): 4.
86. Labor Union Archives, Prague (henceforth: LUAP), Staré odborové spolky, sign. 4384/317, Zápisy z jednání představenstva Svazu českých kovodělníků, zápis z 29. července 1914.
87. *Právo Lidu* (August 15, 1914): 5.
88. *Právo Lidu* (August 31, 1914): 2.
89. *Kovodělník* 35 (August 20, 1914): 142.
90. *Nová Doba* (August 20, 1914): 3.
91. *Večerník Práva Lidu* (September 2, 1914): 4.
92. ACCP, Referát XIV. Zásobování, inv. č. 29, Protokoly aprovisační komise za rok 1914, protokol z 1. září 1914.
93. *Vznik a činnost kuchyní komitétu pro společné stravování méně majetného obyvatelstva v Praze v létech 1916–1920* (Prague, 1920), 1–2.
94. Jakub Rákosník, *Odvrácená tvář meziválečné prosperity. Nezaměstnanost v Československu v letech 1918–1938* (Prague, 2008), 69–84.
95. Příloha *Práva Lidu* (December 24, 1916): 1.
96. Ibid.
97. ACCP, Referát XIV. Zásobování, inv. č. 33, Protokoly ředitelství aprovisačních ústavů za rok 1918, protokol ze 4. června 1918.
98. ACCP, Referát XIV. Zásobování, inv. č. 31, Protokoly aprovisační komise za rok 1916, protokol z 31. října 1916.
99. Večerník *Práva Lidu* 183 (August 13, 1917): 1.

100. Večerník *Práva Lidu* (July 20, 1917): 2.
101. *Nová Doba* (July 7, 1917): 2.
102. *Právo Lidu* (September 7, 1914): 3. On the history pf the Prague slaughterhouse during World War I, see Miroslav Moutvic, *Ústřední jatky města Prahy v Holešovicích: 1895–1951* (Prague, 2007), 24–30.
103. *Vznik a činnost kuchyní komitétu pro společné stravování méně majetného obyvatelstva v Praze v létech 1916–1920* (Prague, 1920), 2.
104. ACCP, Referát XIV. Zásobování, inv. č. 31, Protokoly aprovisační komise za rok 1916, protokol ze 4. července 1916.
105. ACCP, Fond Ústřední sociální úřad (POM: 152), sign. IV. 1a, inv. č. 333, zápis ze schůze evidenční komise pro stravování z 22. 3. 1918.
106. Mesch, *Arbeiterexistenz in der Spätgründerzeit*, 121–53.
107. The duty to bring their own knives and forks was explicitly mentioned in the respective rules for each of the public kitchens. ACCP, Referát XIV. Zásobování, inv. č. 29, Protokoly aprovisační komise za rok 1914, protokol z 2. října 1914.
108. The presence of policemen in the first public dinnery was introduced practically immediately after its opening. Newly opened dinneries were equipped with policemen from their very start. ACCP, Referát XIV. Zásobování, inv. č. 29, Protokoly aprovisační komise za rok 1914, protokol z 27. října 1914.
109. *Vznik a činnost kuchyní komitétu*, 14.
110. ACCP, Referát XIV. Zásobování, inv. č. 29, Protokoly aprovisační komise za rok 1914, protokol z 2. října 1914.
111. ACCP, Referát XIV. Zásobování, inv. č. 29, Protokoly aprovisační komise za rok 1914, protokol z 6. října 1914.
112. ACCP, Referát XIV. Zásobování, inv. č. 35, Aprovisační protokoly obce Branické od 17. dubna 1915, protokol z 12. června 1917.
113. ACCP, Fond Ústřední sociální úřad (POM: 152), sign. IV. 1a, inv. č. 333, zápis ze schůze evidenční komise pro stravování z (October 20, 1916).
114. Ibid.
115. ACCP, Referát XIV. Zásobování, inv. č. 35, Aprovisační protokoly obce Branické od 17. dubna 1915, protokol z 22. června 1917.
116. ACCP, Referát XIV. Zásobování, inv. č. 31, Protokoly aprovisační komise za rok 1916, protokol ze 4. července 1916.
117. Ibid.
118. Ibid.
119. *Vznik a činnost kuchyní komitétu*, 10–11.
120. Ibid., 11.
121. Ibid.
122. Ibid., 15–16.
123. ACCP, Referát XIV. Zásobování, inv. č. 32, Protokoly aprovisační komise za rok 1917, protokol z 6. února 1917.
124. ACCP, Referát XIV. Zásobování, inv. č. 32, Protokoly aprovisační komise za rok 1917, protokol z 9. ledna 1917.
125. *Vznik a činnost kuchyní komitétu*, 8.

126. Ibid., 25.
127. Ibid., 5.
128. Ibid.
129. *Bericht der k.k. Gewerbe-Inspektoren über ihre Amtstätigkeit im Jahre 1914* (Vienna, 1915), 180.
130. See Šedivý, *Češi, české země a velká válka*, 223–27.
131. *Bericht der k.k. Gewerbe-Inspektoren*, 236–37.
132. Ibid., 407.
133. Ibid., 455.
134. Ulrike Thoms, "Essen in der Arbeitswelt. Das betriebliche Kantinenwesen seit seiner Entstehung," in *Die Revolution am Esstisch*, ed. Hans Jürgen Teutenberg (Stuttgart, 2004), 203–18.
135. František Janáček, *Největší zbrojovka monarchie. Škodovka v dějinách, dějiny ve Škodovce 1859–1918* (Prague, 1990), 397; Gustav Habrman, *Mé vzpomínky z války. Črty a obrázky o událostech a zápasech za svobodu a samostatnost* (Prague, 1928), 132–33.
136. On the iner structure of the enterprise see Václav Jíša, *Škodovy závody 1859–1919* (Prague, 1965), 289–402.
137. Nedvěd, *Jak to bylo na českém západě 1914–1918*, 161–62.
138. Central Military Archives in Prague (henceforth: CMAP), fond 9. Sborové velitelství (1883–1919), karton 96, č. res. 39445. Hromadný dopis správám militarizovaných podniků v Čechách ze 13. září 1916.
139. CMAP, fond 9. Sborové velitelství (1883–1919), karton 96, č. res. 44376. Opakovaný hromadný dopis správám militarizovaných podniků v Čechách z 15. října 1916.
140. CMAP, fond 9. Sborové velitelství (1883–1919), karton 96, č. res. 54310. Hlášení z Teplic—Sobědruhů z 10. listopadu 1916.
141. CMAP, fond 9. Sborové velitelství (1883–1919), karton 96, č. res. 53165. Hlášení z Mostu ze 26. listopadu 1916.
142. Ibid. Hlášení z Turnova z 3. prosince 1916.
143. Loewenfeld-Russ, *Die Regelung der Volksernährung im Kriege*, 65–67.
144. CMAP, fond 9. Sborové velitelství (1883–1919), karton 96, č. res. 5913. Hlášení z Mostu ze 4. února 1917.
145. CMAP, fond 9. Sborové velitelství (1883–1919), karton 96, č. res. 7204. Hlášení z Mostu z 13. února 1917.
146. CMAP, fond 9. Sborové velitelství (1883–1919), karton 96, č. res. 11980. Hlášení z Mostu z 15. března 1917.
147. CMAP, fond 9. Sborové velitelství (1883–1919), karton 96, č. res. 14147. Hlášení z Mostu z 27. března 1917.
148. CMAP, fond 9. Sborové velitelství (1883–1919), karton 96, č. res. 18399. Hlášení z Mostu z 22. dubna 1917.
149. CMAP, fond 9. Sborové velitelství (1883–1919), karton 96, č. res. 23713. Hlášení z Teplic z 31. května 1917.
150. "Vojenské velitelství v Praze císařské vojenské kanceláři v Bádenu. Hlášení o zásobovacích poměrech, smýšlení a hospodářské situaci dělnictva v Čechách

ze 14. března 1917," in *Sborník dokumentů k vnitřnímu vývoji českých zemí za 1. světové války*, ed. Alexandra Špiritová (Prague, 1996), vol. IV., document no. 13, 55.

151. ACCP, Referát XIV. Zásobování, inv. č. 31, Protokoly aprovisační komise za rok 1916, protokol z 10. října 1916.
152. On the Czech case, see Jana Machačová and Jiří Matějček, "Chudé (dolní) vrstvy společnosti českých zemí v 19. století. Sociální pozice a vzory chování," in *Studie k sociálním dějinám* 1 (1998): 121–303; Machačová and Matějček, *Nástin sociálního vývoje českých zemí 1780–1914*, 138–89; Fasora, *Dělník a měšťan*, 35–259; Jiří Matějček, "Dělnické hnutí v Českých zemích do roku 1914. Emancipace dělnictva, nebo hegemonie proletariátu? Pokus o objektivní hodnocení vývoje hnutí i stavu výzkumu," *Studie k sociálním dějinám* 2 (1998): 153–95; Jana Englová, "Dělnictvo jako subjekt a objekt historického bádání," in *Problematika dělnictva v 19. a 20. století. Bilance a výhledy studia*, eds. Stanislav Knob and Tomáš Rucki (Ostrava, 2011), 34–39; Stanislav Knob, "Stávkové hnutí v československé i zahraniční historiografii, srovnání a výhledy," in *Problematika dělnictva v 19. a 20. století. Bilance a výhledy studia*, eds. Stanislav Knob and Tomáš Rucki (Ostrava, 2011), 65–72.
153. Tanner, *Fabrikmahlzeit*; Victoria De Grazia, ed., *The Sex of Things: Gender and Consumption in Historical Perspective* (Berkeley, 1996); Monica Neve, *Sold! Advertising and the Bourgeois Female Consumer in Munich 1900–1914* (Stuttgart, 2010); Martin Daunton and Matthew Dilton, eds., *The Politics of Consumption: Material Culture in Europe and America* (New York and Oxford, 2001).

Chapter 2

Rationed Fatigue: The Politics of Work

"A brave horse dies in its harness."
—Otto von Bismarck[1]

Iron in the Main Role

The camera of a silent film slowly pans from right to left inside a factory courtyard. Human forms occasionally appear around heaps of iron, trying to organize the amorphous iron piles. The camera pays them no attention and continues on until it stops several meters above the iron tracks that intersect the courtyard. It zooms in. Along the sides of the tracks, more human figures move about, pushing carts loaded with heavy iron. Cut. A new shot shows the mechanized cable that uses two opposing carts to help push the piles of iron through the jagged factory space. Cut. The camera pans over the factory courtyard once more. A train full of coal rides into the middle of the shot. Sixteen carts fit into the shot, but the train no doubt has more. On the left side, there are dozens of cowering figures in skirts and pants, shoveling the dumped coal into piles. The subtitle informs us that one such delivery earns the factory six hours of production.

Several shots forward, the camera is aimed at the film's main subject—the manufacturing processes at the factory. The camera is placed above a row of furnaces, into which mechanized machines are adding iron. After removing the iron from the flames, a figure appears from the darkness and checks the iron rods. Other shots continue to track the paths of the raw iron through the steelworks, but their basic composition remains the same. The focus is always on the iron transforming itself through heat, emitting an almost blinding light. Everything else is in shadow. The furnaces, tracks, cars, electromagnetic cranes, and people enter into the frame only when they are actively participating in the long, technically and physically arduous transformation of raw iron into high-grade steel. The main star of the film, which the camera constantly follows, is iron and its transformation into steel.

When one of the human figures comes into the frame, it is only to complement the overall composition. The rods, which the founders use to check the melting of the iron, are organized into geometric shapes, and the workers who use them usually move about in a very small space so as not to disturb the overall layout of the scene. A short cut-in shows several carts of the iron train, in which steel castings are neatly arranged in rows of four. The rows are always placed crosswise. Then there is a shot of the factory courtyard, which puts the previous shot in context. The entire courtyard, several hundred square meters wide, is covered in regularly arranged castings, again in rows of four.

FIGURE 2.1. Iron in the Main Role: Das Stahlwerk der Poldihütte, part 1.

The figures emerging from the shadows shorten the individual iron rods, pieces of which appear in the other factory halls, this time at a much more delicate counter. Standing next to it are two figures—a man and a woman—who, using various tools, turn a steel rod on a machine tool counter and neatly lay the processed piece down several meters away. Behind them the whole time stand two men in suits, closely observing their every move. In the meantime, in the other part of the hall, a group of men is loading the processed steel moldings onto movable carts. Each of the moldings ends up in the hands of a woman, who loads it into another furnace aided by two strong men. After several short moments, the molding is taken out and the two strong men use a pulley to place it on the counter. Two men in hats wearing bow ties around their necks, smoking cigarettes, stand next to them and once again closely observe their work. Roughly thirteen minutes into the film, the viewer begins to slowly understand where the path of the brightly shining pieces of iron and steel leads. After the pieces of steel go through several furnaces, they take on the shape of artillery grenades, which the workers load onto carts and manually transport to the other factory halls.

A cut and a shot of a factory courtyard follow. Here, the shells of the artillery grenades are arranged into endless rows of columns. At the edge of the frame, six women are pushing a train full of grenade shells along a track. The next shot shows several older children, who are also trying to help assemble the munitions pieces into endless regular rows. Another much longer take shows the lathe hall, where workers carve grooves onto the grenades. Carts ferry the unfinished grenades down a track to the individual lathes. At each stands a man, woman or adolescent, who works the grenades piece by piece on the lathe. The caption indicates that there are five hundred such lathes and fifteen hundred people in total work here.

The next shot of the table, where the finished grenades are precisely measured, is framed with a background dominated by pyramids of final products. The bottom surface of the pyramid is always a square and grenades are stacked upon it. An identical square is placed on top of it, made out of the same amount of grenades minus one. On top of this square there is another square with one grenade less on either side until the pyramid ends on the last level. The pyramids stand next to one another in a neat row. The next shot reveals that they fill up the entire storage hall. There are regular gaps between the pyramids and the entire storage hall is thus a display of impressive symmetrical geometric shapes.

The entire second half of the film focuses on the motif of perfect organization. All stages of grenade production are characterized by the arrangement of the production and the products themselves into graceful, geometric shapes. Every part, every machine and each human figure has

FIGURE 2.2. Triumph of Geometry: Grenade Storage in the Poldi Steelworks, Das Stahlwerk der Poldihütte, part 2

its predetermined place, and together they create a symphony of regular forms and processes. Everything is neatly and efficiently organized to make sure that the mechanical operations and manual labor are done as quickly, effectively and rationally as possible.

The final part of the film turns its focus from the artillery grenades to the technically even more delicate production of crankshafts for airplane and automobile motors. Steel, in various stages of manufacture, once again plays the main role: from a shapeless block to a precisely crafted shaft. In the middle of the shot, a crude steel block appears, which is placed onto a high tonnage press. From the right and left edge of the shot, rods are aimed at the block, holding it in place so that it can be precisely pressed. After a while, the men holding the rods flit past. The rest remain out of the camera's range, making it appear that the steel block is holding itself in the press. Just like in the first part, the last part is dominated by shots of the brightly shining pieces of steel, casting a shadow on everything else. Workers emerge from the shadows only when the steel rods need to be held down or gently shaped; the individual human shapes only briefly pass through the shot and are indistinguishable from one another.

The final products are filmed very differently. In several long shots the camera moves past a row of more than ten crankshafts of various shapes and sizes. The viewer has enough time to differentiate between one model and another, and to think about which shaft is for airplanes and which is for trucks. The final shots present some of the other factory products as well. Whenever possible, the human figures are not in the shot at all. The flexibility of a bridge-supporting spring is shown with a focused shot of the spring itself, which is mechanically stressed by an ingenious bending machine. A human figure steps into the frame at the very end to briefly place a meter on the cable and check the prescribed level of flexibility and firmness. The film ends with a scene where a group of workers, their backs to the camera, are pushing a thin strip of steel through a rolling machine. The steel strip shines brightly, hiding what little of the human faces the camera could capture during the rapid movements of the working bodies.

The first and the third part of the film have a similar emphasis on scenic composition, which always centers on the processed material. The film tells the story of the steel transformed into various end products through different production processes. Iron, from which steel is created, eventually leading to the production of grenades, shafts, propellers and other products designated for front-line combat, plays the main role. Everything happens elegantly, effectively and in constant repetition. The viewer is carried away by the aesthetic celebration of modernity, occasionally complemented with human bodies. The film's portrayal of the production process allows one to forget that any human strain was involved or that the human bodies were in constant danger of serious injury due to the continual presence of material over 1,000 degrees Celsius, or due to their close proximity to several-ton presses.

Dream of Endless Work

The film *Das Stahlwerk der Poldihütte* (The Poldi Steelworks), the subject of the preceding pages, is one of the finest works of wartime Austrian cinematography.[2] It was filmed in 1916 by the "Österreichisch-Ungarische Sascha-Messter-Filmfabrik GmbH" Film Company based in Vienna. The biggest film magnate of the time, the German producer Oskar Messter, also participated in its creation.[3] With a length of over thirty minutes and because it was a documentary, it stood out from the multitude of wartime motion pictures that various production companies produced between 1914 and 1918 in the Habsburg monarchy. The aesthetic celebration of the maximum rationalization of production in the largest Czech steelworks puts it alongside the top films of German cinematography of the time, such as *Stahlfabrik Krupp* (The Krupp Steelworks) or *Thomaswerk* (The Thomas Works).[4]

This documentary also provides an extremely valuable insight into the transformation of manual labor, which changed the social world of practically all Austrian workers between the years 1914 and 1918. The emphasis that the film places on the maximum effectiveness of the work processes in one of the most important enterprises in Austrian heavy industry,[5] and on the composition of particular takes (the material always significantly lit and the workers in secondary roles) illustrates the wartime transformation of manual labor in most Austrian wartime factories.

In the previous chapter, I focused primarily on the development of the "politics of food," and how it emerged from a specific academic discourse that linked questions on the distribution and consumption of food with the issues of the working human body. Within this discourse, food became the necessary fuel for human motors, without which wartime production would not be possible. Focusing now on manual labor reveals that it is the flipside of the same coin. The metaphor of the human motor, which before World War I became the dominant conceptual framework for thinking about the working human body, not only addressed questions about the working-class diet, but also questions on the nature of manual labor and its role in the war effort. In the working world, human bodies became indistinguishable from modern motors and other pieces of factory equipment. As the camera revealed in "The Poldi Steelworks," they served—like combustion engines, lathes or the large furnaces—as material machines and therefore were merely the context in which the all-important final product was created.

"The Poldi Steelworks" was thus a part of the discourse on manual labor that established limits within which it was possible to talk and think about work, ultimately determining what could be considered work and under what conditions it could occur. The influence of this discourse on workers' everyday experiences was enormous. In order to uncover its limits, we must, as in the case of food, first examine the decades before 1914. The understanding of man as an independent type of motor was firmly established in European thought in the last third of the nineteenth century, not only in terms of his relationship with food as a source of energy, but also in his relationship to work as the main result of the human body's activity.

If the human body worked like a motor, as the science of the time postulated, then a logical step in its relationship to work output was to measure its work capacity and performance as precisely as possible. As the biggest obstacle on the path toward the maximum effectiveness of industrial production, human fatigue was a phenomenon that quickly became the center of scientific research on manual labor.[6] In the decade before World War I, various experiments were conducted to figure out how to eliminate it. The context of World War I, as in the case of food, significantly widened the possibilities of applying prewar academic knowledge to human labor and fatigue, thus influencing the everyday experience of millions of workers in Austro-Hungarian factories.

Fatigue as a phenomenon that could prevent human bodies from working uninterruptedly was part of the complex of problems that industrializing societies had been dealing with since the 1830s. The proletarianization of factory workers was connected with many negatively viewed social phenomena, such as hazardous living conditions, an increase in diseases and mortality, and destabilized family structures, which created a complex of problems that gradually came to be called the "social question." German-language journalism and politics especially had been using this term with greater frequency since the 1830s and the revolutions of 1848–1849 were seen by their supporters and opponents alike as events that were partly triggered by accumulated social problems.[7]

Various proposals to solve the "social question" had been occurring in Germany and Austria with increasing intensity since the 1870s. These combined scientific approaches emphasized the necessity to deeply research the social problems, measure them statistically and subsequently consider a solution. Their language also had a moralizing character, which did not see the solution to the social question in the mere improvement of material conditions of the industrial poor, but in deeper moral reforms.[8]

After 1880, and even more so after 1890, the solutions to the "social question" increasingly used scientific language, i.e., used statistical calculations and targeted government measures, which began to be implemented in Austria during the 1880s.[9] These were not primarily motivated by humanitarian concerns, but represented an effort to describe and assess the suffering of the industrial poor, and subject it to complex planning with the aim of increasing economic productivity. The seemingly foreign world of industrial poverty had to be made understandable to the middle class through objective calculations, statistics, or textualized descriptions. The social conditions of industrial workers were translated into the language of symbols and numbers, which subsequently allowed middle-class reformers to solve the "social question" alone, disconnected from the working-class environment.

Modern science could thus authoritatively enter into the debates on the solutions to the "social question" and offer what it considered to be neutral and objective solutions to the economic and social conflicts brought about by industrialization. Moral reformation was gradually replaced by statistical analysis and the distance between industrial workers and the middle class was legitimized not by "God's Order" based on Christian theology, but by the objective, scientifically justified rationality of market mechanisms, which determine the social position of all individuals based on their own efforts.[10]

The crowning triumph of science in the struggle for primacy in the solution to the "social question" happened in 1911, when Taylor's *Principles of Scientific Management* was first published in the United States.[11] The founding work of modern management, it quickly became a worldwide bestseller. It emphasized the maximum rationalization of factory operations, which concerned not only each employee, but also the minutest task; the slightest human movement had a precisely determined time and place, confirming the primacy of scientific discourse over the moral one. Many pre-1914 social reforms reflected this positivist scientific discourse, which was able to authoritatively establish the optimal conditions of industrial production based on objective scientific findings. Indeed, one of Taylor's main justifications for his research was that he would be able to determine each worker's fair amount of work and, based on this amount of work, he would finally establish a fair salary. All political disputes between labor and capital would be resolved forever with objective scientific knowledge.[12]

The basic characteristic of the new scientific management of work was, as in the case of food, the understanding of manual labor as an activity that was carried out primarily by the human body. The human body and its physical disposition toward exertion and fatigue was the decisive object of research. The phenomenon of human fatigue, however, was an obstacle to the attainment of the modern utopia of maximum effectiveness and, as such, occupied the very center of scientific research. Human fatigue was seen as the main obstacle to work, gradually replacing the older concept of laziness, an attribute harshly denounced by Christian morality in particular, which lost much of its social relevance by the end of the nineteenth century. Large factory operations did not need Christian theology to impose discipline from the outside, but an internally regulated human body, which would, like a machine, automatically fulfill its work responsibilities with no recourse to any moral imperatives. The ideal of a worker who was led by moral authority or constantly monitored by the continual supervision of his superiors thus gradually faded and the ideal of a working body directed by its own inner laws—the laws of the human motor—came to the fore.[13] The relationship between the human body and work was not expressed in the morally coded categories of diligence and idleness, but in seemingly objective scientific categories of human work capacity and fatigue. Modern man did not cease working because he was lazy, and thus immoral, but because the objective phenomenon of fatigue did not allow him to continue. If physical fatigue was sufficiently understood and analyzed, according to the leading opinion, it would open up new possibilities to completely eliminate it.[14]

Like food, the phenomenon of fatigue was subsumed into the discourse of modern science, primarily of physiology, but of other sciences to a certain extent as well. As it did with the other objects of its research, modern science objectively analyzed fatigue, testing it in a laboratory and finally quantifying it. In the decades leading up to World War I, a large network of European and American research facilities was created to analyze how fatigue originates and what causes it, and formulate protocols for its prevention through objective measurements and experiments.

At the end of the nineteenth century, physiologists first began to improve the methods for measuring and representing human fatigue. The year 1884 was a turning point in this endeavor, when the Italian Angelo Mosso constructed the first ergograph—a device that measured the length and intensity of muscle contractions with mechanical or electric sensors during a specific stress test, such as, for example, lifting weights or pushing down on springs. During the experiments, the research object's muscle was attached to the device and stressed until total exhaustion. The device displayed the length of the contractions, their frequency and showed the muscle activity in intervals of time.[15]

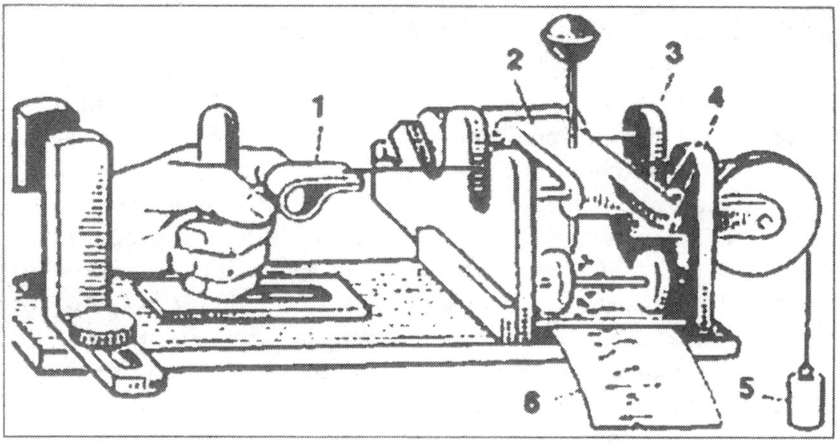

FIGURE 2.3. Mosso's Ergograph for Stress Testing the Index Finger
1. Movement Sensor
2. Recording Unit
3. Holder
4. Unit Moving the Strip Graph
5. Weight
6. Strip Graph Showing Muscle Activity

Mosso made hundreds of measurements of dozens of human bodies with the ergograph and summarized his experiments in 1891 in a key work simply titled *Fatigue* (La Fatica).[16] The book quickly became a European sensation and turned its author into one of the main representatives of the avant-garde of European physiology. However, the results of his experiments were initially ambiguous. Mosso's research subjects, hooked up to the ergograph, became fatigued in various ways and the quantified results of his experiments did not point to any kind of connection between muscle fatigue and gender, age or occupation. Some of the research participants became fatigued suddenly and quickly, while others experienced muscle fatigue slowly and gradually. The regularity in the onset of muscle fatigue first occurred when specific individuals were hooked up to the ergograph repeatedly. While different people experienced fatigue at varying stages of physical exertion, an individual usually experienced it at the same point. At the same time, the objectively measured fatigue did not always correspond with the subjective feelings of the measured individual. During the experiments, some people stated that they were fatigued long before their muscle activity began to weaken; others did not feel fatigued even when the tested muscles were not performing at their initial strength.[17] Based on his observations, Mosso concluded that fatigue is an objective phenomenon that has nothing to do with a person's will, but is subjugated only to the physiological laws of the human body.

For Mosso, fatigue was thus a type of disease, the onset of which was influenced by a person's physical constitution, lifestyle, and other factors. He tried to prove this thesis in his world-famous experiment, in which he gave a relaxed dog a transfusion of blood from another dog, which had been exercised to exhaustion. Signs of exhaustion in the originally relaxed dog allegedly could be seen immediately.[18]

As the origin and incidence of fatigue became better understood, around the year 1900 scientists began researching methods and procedures to eliminate it. Following Mosso's famous experiment with the dogs, the first direction that this research took was to try to treat fatigue clinically. These efforts culminated in 1904, when the Erlangen professor of physiology Wilhelm Weichardt announced that he had invented a "fatigue vaccine." The idea that fatigue was caused by specific toxins produced by the human body during physical exertion led the German professor to develop a vaccine that would stop the production of so-called "kenotoxins."[19]

Aside from this clinical approach, around 1900 specialized scientific experiments took place in which researchers tried to develop complex methods of overall fatigue prevention. The rhythm of bodily movements, their frequency, intensity and range were subjected to detailed laboratory experiments, which were supposed to propose each muscle's optimum movements in order to diagnose fatigue in time and then appropriately plan out work assignments and their length. Such planning would ensure that the working body would avoid getting fatigued. In 1894, the first "chronophotographs" of workers were presented in Paris, capturing a worker's body banging a hammer on an anvil in several quick sequential snapshots. Thus there was a picture of work being performed, divided into minimal segments of time, with which it was possible to explore the most effective movements of all of the important human body parts and propose their optimization so that they would consume the least amount of energy.[20]

The research on human fatigue as an academic discipline was a very lively, dynamic, and international field at the end of the nineteenth and beginning of the twentieth century. Large congresses, in 1900 in Paris, in 1903 in Brussels, in 1907 in Berlin, and in 1912 in Washington, contributed to the sharing of the newest findings, produced primarily in French, German, and Italian laboratories around the industrial world. The translation and distribution of seminal works into various European languages, especially into French, i.e., the main language of the European scientific community, and their distribution across Europe also helped.[21]

The main work from the leading academics of this field, such as Charles Richet, Josefa Ioteyko, and Ferdinand Lagrange, were written in French directly or translated into French quickly, so they arrived in the main scientific libraries of Europe, including those in Vienna and Prague, with minimal delay.[22] Similarly, Mosso's basic work was accessible to the academic community of Austro-Hunagrian physiologists in its German translation, which was published just one year after the original Italian edition.[23] In the Czech environment, Mosso's work was further known due to his treatise on the best physical education practices for teenagers, published in 1894 and translated into Czech in 1901, which attempted to apply the results of his research on physical exercise in school.[24]

Next to the French studies and Angelo Mosso's main works, original research from imperial Germany, in which German academics advanced discoveries about the physical aspects of human movements at work, also played a role in the Austrian scholarly community. Their work was received more intensely in the German-speaking scientific circles of the Austro-Hungarian Empire than French works, and they were part of the fast and intense transfer of the latest scientific findings on human labor and fatigue into the Habsburg Empire. Similarly to the French experiments of Charles Richet or Ferdinand Lagrange, books by the leading German physiologists became well-known very quickly in Vienna and Prague.[25] The university textbooks of the renowned German physiologists Nathan Zuntz became one of the main physiology textbooks in Austrian universities before World War I. Similarly popular were some of Emil Kraepelin's works, which focused on the influence of various factors on the "work curve" of the human body or on the phenomenon of fatigue stemming not from manual, but from spiritual labor.[26]

Austro-Hungarian army research in turn focused on Weichardt's anti-fatigue vaccine, and it was Austrian army medics who definitively refuted his discovery of the so-called kenotoxins as special fatigue-causing toxins. Broad experiments conducted by the Habsburg army on its soldiers prior to, and at the start of, World War I largely concluded that the application of Weichardt's prophylaxis did not prevent the onset of fatigue. On the contrary: the results were extremely disappointing, especially in comparison with a new method of increasing productivity—the intravenous application of caffeine concentrate.[27]

As in other industrial parts of Europe, shortly before 1914 a scientific discourse was established in the Bohemian lands that conceived of the body as a mechanical converter of energy not only in terms of food, but also in terms of carrying out various jobs. Czech physiology saw "life energetics" as the main subject of its research,[28] i.e., the principles by which the body manages energy and which limits its capacities for work. The further study of the function of human muscles and conditions of the onset and course of fatigue led to a generally accepted conclusion within the Czech scientific community that "intervals of rest between periods of activity are a condition for maintaining the organism's sensitivity. Muscle recuperation, marked by a return to sensitivity, is contingent upon rest."[29]

The results of this physiological research immediately began to influence the social environment of industrial labor. Demands accumulated for state regulations of the labor code, which would take into account these newest scientific findings and increase economic productivity with a suitable modification of the rhythm of work. Workers' muscles needed respite in order to generate sufficient outputs. The adjustment of working hours before 1914, the foundations of medical and accident insurance or the prescribing of precisely scheduled work breaks mirrored this scientific discourse and were often understood not as concessions to workers' demands, but as a targeted intervention that would increase work effectiveness.[30] "We don't see medical insurance merely as a humanitarian and moral command. The fact that it is one of the most effective preventive measures to raise health standards and thus to preserve the workers' ability to work also makes it one of the most important questions of economic effectiveness,"[31] argued Young Czech representative Karel Adámek in 1887 during the parliamentary debate on implementing workers' medical insurance in Austria.

These broad social reforms put pressure on the renewal of the work force and the rhythmization of working hours, which was to be divided into regular segments of physical exertion and breaks, during which the human body's capacities would regenerate. If anything was missed during the break, it was then caught up by working more intensely in the next shift. Exertion and rest was precisely scheduled in order to forestall the onset of fatigue. This led to the more intensive "economization of time," i.e., the submission of working bodies to the economic rationality of maximum effectiveness, which exactly described the activity that the working body had to do at every moment, just like with inanimate motors. As Josef Pazourek, one of the Czech theorists of work organization, and later the rector of the Czech Technical University, summarized on the eve of World War I, human labor and its individual activities should be "conducted most methodically and economically, in order to eliminate useless movements, replace difficult movements with easier ones, looser movements with faster ones, less effective movements with more effective ones, in order for the worker to become most efficient."[32]

The outbreak of World War I thus provided unforeseen possibilities for the rationalization of work to become the main organizational principle of a substantial part of the Austrian industry. It was necessary to get the human motors to work as hard as possible; the Austrian wartime economy literally became a laboratory of effectiveness between the years 1914 and 1918.[33] As was the case with food, manual labor became part of a complex discourse structured around the perception of the human body as a motor. When the destruction of the enemy became Austrian industry's only goal in the fall of 1914, it was necessary to mobilize all available sources of energy and this mobilization also pertained to human motors. As Jindřich Fleischner, one of the leading Czech theorists of the organization of labor, wrote in 1915: "A human can achieve a higher level of his motor's effectiveness, but his work must be better utilized if it is to become the means to higher goals."[34] During the war years there was no "higher goal" than the defeat of the reviled enemy and thus human motors were fully subordinate to the Austrian war effort.

Factory Equipment

The social environment of industrial labor in the wartime Habsburg Monarchy was legally regulated by several fundamental laws that defined relationships between employers and employees. The most important one was the "Law on Wartime Operations" (*Kriegsleistungsgesetz*) no. 236, passed on December 26, 1912 during the Balkan Wars.[35] This law enabled the government to subordinate large parts of civil life to the demands of the army and wartime production in the event of a declaration of war. Any working man between the ages of 18 and 50 (later 52) could be, according to this law, drafted against his will into wartime production, subject to martial law and stripped of his basic civil rights.[36] He was then firmly bound by this legal relationship to his newly assigned workplace, which abided by military discipline and was subject to military justice.[37]

Martial law, to which the enterprises were subject, draconically punished any signs of association and resistance. The employers had judicial powers—they could freely sentence people to up to 15 days in jail. For more serious crimes, workers could be convicted by a military court and sentenced to up to 20 years in jail, or to death. Military courts, moreover, repressively interpreted the relevant law and harshly punished any signs of organized labor.[38]

In July 1914, law no. 155 on state protected enterprises was passed, making it possible to prosecute any refusal to work.[39] The official wording of the relevant paragraph that allowed anyone who "fully or partially refuses or neglects the fulfillment of his work in such a way as to hinder the service or the enterprise"[40] to be sentenced to jail made practically anything punishable by law. The far-reaching application of this law during the war years deprived the workers of any possibility of legal collective negotiations.[41] This legal amendment went through certain changes in March 1917, when so-called complaint committees were established where workers could submit their grievances. Their authority, however, was very limited and the basic, disciplinary framework shaped the world of industrial labor until the fall of 1918.[42]

Overall, labor relations were removed from the authority of public law and to a great extent were placed under the jurisdiction of criminal law. Immediately after the outbreak of war, many enterprises in the Bohemian lands came under the liability of these two laws, which changed the working conditions of hundreds of thousands of workers practically overnight.[43] Some enterprises went even further than the relevant regulations, and the order that they imposed upon their workers was even stricter. The Prague Ringhoffer factories, for example, did not just prosecute "the refusal and neglect of the fulfillment of work responsibilities," as the law stated, but explicitly punished any "disobedience toward superiors," while each superior himself defined what constituted disobedience.[44]

The far-reaching mobilization for total war in many places meant putting the originally only academically conceived metaphor of the human motor into practice. The mobilization effort of the state and the scientific concept of a rationalized worker's body blended into one set of practices. Workers were in many respects seen as specific factory equipment, which could be freely disposed of just like any other tool or machine. Austrian state and private employers, for instance in connection to the stationing of workers in various areas of the wartime industry, diligently used the same vocabulary that they used to describe the usage of other factory equipment. Factory workers were thus "used" and "utilized" (verwendet), just like machines they "performed" (verrichten) their work and in places where there were not enough workers, their "higher demand" (wachsender Bedarf) was declared.[45] Their work rhythm was generally precisely determined by the internal order of the factory and regulated by the sound of the factory siren, bell, or other such device that signaled the beginning and end of individual work shifts and set in motion or calmed down machines as well as working bodies.[46] In the Pilsen Škoda factory, for example, all workers were even forbidden from washing their hands before the clanging of the bell.[47]

This objectification of workers' bodies into tools of production had many facets, which publicly demonstrated the utter transformation of the prewar order. At the very beginning of 1915, the military administrator of the Pilsen Škoda factory, Kühn, built a special workers' jail on the grounds of the factory in order to ensure maximum effectiveness not only of the work, but also of the administration of punishments for violating work responsibilities.[48] Transporting workers who were punished with several days in jail out of the factory proved to be too complicated, for the offender had to be accompanied. Transport to the Pilsen prison and subsequently back to the workplace was, moreover, a useless waste of time and workforce and a jail inside the factory compound removed these shortcomings and was an efficient use of manpower. The punished workers' journey to the new jail only took several minutes and the offenders could return to work immediately after serving their sentence.[49] The following year, convicted workers spent a total of 11,556 days in jail, so the jail was almost certainly never empty, and for twelve months every third worker spent one day in jail on average.[50] The administrator of the Poldi factory in Kladno, František Hložek, implemented a similar arrangement, ordering a special jail for punished workers to be built in one of the factory courtyards. The only difference from Pilsen was that workers' sentences were divided into eight to twelve-hour-long segments, so that workers could serve out their sentences during the night. During the day they could work and then return to spend the night in jail.[51]

Many employers did not just use their penal authority for jail sentences, but they also publicly manifested their superior position in the workplace and their power to freely handle workers' bodies. The world of industrial labor experienced an unprecedented invasion of physical violence that had previously been solely connected to the environment of the Austro-

Hungarian army. Punishments in the form of thrashings were often carried out and it was not rare for workers to be strung up with their hands tied behind their backs with their feet barely touching the ground. For the punishments to be effective, if the worker fainted he was revived with a bucket of ice-cold water, ensuring that the punishment could be carried out until the very end.[52] For example, in the Pilsen Škoda munitions plant from October 12, 1915 to May 12, 1916, i.e., a period of seven months, there was a total of 31 hours of this kind of physical punishment.[53] Similar physical punishments, unlike depriving workers of their personal freedoms through jail sentences, had a significant public-symbolic component. Thrashings as well as hanging by hands were usually carried out in the public space of the factory, before the eyes of male colleagues, female employees and children.

The military administrators were not only authorized to demonstrate their competencies most obviously, but they were also called upon to do so by the central military offices. According to instructions that were issued in May 1916 by the command of the 9th army corps in Litoměřice, which administered all of the enterprises in northern and northwestern Bohemia, all military administrators frequently had to notify workers "of the harsh punishments that would follow immediately, had the delivery of products to the various enterprises been prevented."[54] Many administrators took such statements to heart and often took advantage of the opportunity to publicly demonstrate their unlimited competencies to handle workers' bodies. In Pilsen, the local military administration not only ordered workers to be hung up by their hands or publicly thrashed by an authorized person, but also ordered special punishments as well, such as when the thrashing was publicly carried out by family members in front of other employees. Cases, in which a father, for instance, was forced to publicly beat his son to a bloody pulp in the factory courtyard had only one purpose—to widely demonstrate the unwavering and unlimited power of the military administration over workers' bodies.[55]

This transformation of the basic relationships in the workplace did not just affect a quantitatively insignificant group of people. The machine-building industry especially became very important during the war years, when its portion of the overall economic production of the Habsburg Monarchy increased; the absolute number of workers employed in it rapidly increased as well. If we look at the munitions and armament plant in Pilsen, which was the largest enterprise in the whole Habsburg Monarchy, during the first two war years of 1914 and 1915 we can observe a more than 100-percent increase in the overall number of the workforce. While in 1914 the Škoda factories employed 9,364 people, in 1915 there were more than 20,000 people and the number continued to rise.[56]

The amount of enterprises and persons who were subject to special wartime legislation was not only determined by the transfer of many factories under the jurisdiction of the relevant laws at the beginning of the war, but also constantly increased during the war years. At the end of February and the beginning of August 1916, the entire Bohemian lands experienced a huge personnel turnover, when completely new military administrators were assigned to more than two-thirds of militarized enterprises.[57] These administrators were granted new powers, allowing them to control not only the enterprises entrusted to them but also the workers who fell under military jurisdiction in all of the other enterprises in the given district. Military administrators thus gained practically unlimited powers to interfere in the operations of any of the enterprises in the given district, where they could exercise their right to freely dispose of the workers.[58] In the Pilsen Škoda factory in 1916, there were more and more cases of workers who worked in the factory as employed civilians and were quickly drafted into the army, only to be immediately sent back to the same job. They continued working, but were subject to the military justice system and intensified discipline, and they no longer received the same salary, but a much lower soldier's pay.[59]

Just like inanimate machines, many workers were the focus of maximizing the effectiveness of their work and output. Special attention was paid to compliance with prescribed work breaks that were supposed to regenerate workers' necessary physical strength. Throughout the war, emphasis on this aspect of physical work was foremost in the factory managements' interests, as well as those of the state itself. The state had to limit and, at the same time, precisely focus its supervision over the industrial factories under conditions, in which its own bureaucracy was significantly reduced by the draft. For instance, in Pardubice and České Budějovice the first wave of drafts in the fall of 1914 affected the trade inspection to such a degree that the local inspectors did not meet once by the end of the year.[60] In Prague, the number of inspections decreased roughly by a fourth immediately after the outbreak of war, while the protracted state of war made the situation in the next several years even worse.[61] In 1913, there were 42,000 inspections in the entire territory of Austria; in 1916, there were only 18,000.[62]

If we look at the trade inspection's agenda more closely, we can see that compliance with working hours and prescribed breaks played a key role. During their visits to Czech enterprises, inspectors focused on how these enterprises complied with regulations that right before the war implemented the latest scientific findings into everyday factory life. The most important of these regulations was Decree no. 187 from 1912, which stipulated the minimum length of work breaks and strictly established that these breaks must be scheduled throughout the working day so that work shifts would be divided into segments of equal time.[63] The longest allowable work shift was five hours, while the overall working day in factories was not to exceed eleven hours, or ten hours in the case of coal mines.[64] The trade law explicitly forbade the transfer of work breaks to the beginnings or ends of shifts.[65]

As the results of the wartime inspections reveal, large enterprises in particular incorporated these regulations and implemented the prescribed work rhythm. The trade inspectors could thus often conclude with satisfaction that, as far as compliance with the work rhythm was concerned, all of the Austrian and Czech factories maintained a high level of discipline and any violations were concentrated only in several very small enterprises.[66] In the southern Bohemian inspection district of České Budějovice in 1914, only one small factory, in which several employees were forced to work in shifts longer than eighteen hours, violated the work time regulation. Instead of levying a fine, the inspectors elected to issue a reprimand, in which they tried to explain to the factory owner that shortening the workday and complying with the prescribed breaks is first and foremost in the interests of his own production, because overworking the workers can have a negative impact on the overall effectiveness of the work. As an inspector symptomatically summarized: "[T]he managers of the enterprise needed to understand that an eighteen-hour shift needs to be shortened in their own interest."[67]

In the eastern Bohemian district of Pardubice, the inspectors also encountered only one small shortcoming, when in one local laundry several workers cleaned parts of the machinery during breaks that were otherwise observed. Although workers agreed to do this voluntarily and were paid extra for it, the inspection pronounced a vehement interdiction. Non-compliance with the prescribed breaks was not going to be tolerated even when the workers agreed with them.[68] With no consideration as to whether the non-compliance with the working rhythm was imposed by employers or initiated by the workers themselves, it was always considered as a threat to wartime productivity and thus as damaging the interest of the state, which strictly promoted the most effective utilization of the capacities of all human motors. In accordance with the latest scientific findings, these needed to be granted the necessary time to regenerate and cheating it would turn against wartime production and thus against the war effort itself.

In this context, it is not surprising that most of the main wartime production enterprises adhered to the regulations on the precise length of working hours with no problem. Their close connection to the state and army administrations granted them better access to the latest scientific findings on the most effective use of human resources than the smaller factories had. The military administrators and civilian directors of the larger factories were informed about the latest scientific research on human work capacities, the implementation of which promised the factories maximum profits. Those who did not closely follow the current state of the development in management methods or in the study of human fatigue could turn to one of the many "effectivization agencies" that offered to implement the rationalization of production as a "turnkey project." At the end of 1916, for example, two main Bohemian dailies, *Národní Listy* and *Prager Tagblatt*, featured several noticeable half-page advertisements with the following wording: "I will successfully organize your enterprise, I will modernize the running of your factory and organize it according to the most contemporary experiences. I will simplify your factory and make it profitable, and if it is currently profitable, I will make it even more profitable. I will organize the work of your office and factory workers to achieve the highest output that is satisfactory to you and your staff. … I will make the work so systematic that it will bring proof of its rightness at every level. I do everything immediately, only once, but properly."[69]

The advanced rationalization of work, based on contemporary physiological discourse as well as on the emerging field of human resources management, was seen as the crucial step toward maximizing production and thus toward victory in war.[70] The reorganization of labor focused on the division of work into precisely defined and quantifiably objective units of time and labor. In the center of it all was the working body. The factory environment, work conditions, and social relations between employer and employee played lesser roles in considerations of how to increase wartime productivity. The rationalization of production basically equaled the rationalization of the body. Significantly, the success of various rationalization measures depended on the authority of scientific knowledge in industrial societies.[71] These changes in the social world of industrial labor transformed its very foundations, inevitably affecting working-class culture along with the working class itself.

Struggle for Survival: Industrial Rationalization, Working-Class Culture and Turn to the Future

The new structure of work relationships, in which the workers' bodies were equated with tools of production that were at the employer's full disposal, understandably undercut workers' self-identification as socially recognized valuable qualified laborers. In addition to losing the freedom to choose their employer, workers experienced public humiliation in front of their colleagues, women, and children, which degraded their previous relatively high social status.

The first significant change that most workers faced after the outbreak of war was the substantial restructuring of the inner factory hierarchy. The precise definition of work procedures and the time set to do them shifted the main influence on factory operations from paternalistic factory owners, or qualified foremen in workshops, to the new class of professional rationalization experts and managers.[72] They had the necessary expertise to decide when, how, and why to work, which they provided to the factory management either directly as internal experts, or as external consultants.[73]

Hans von Karabatschek, the director of the Pilsen Škoda factory, became probably the most well-known of these experts during World War I. A German-speaking expert in technical sciences and the rationalization of work, he studied in Vienna before the war and then began an illustrious career as a leading Austrian rationalization specialist. Before the war, he worked in several key Austrian heavy industry factories, such as the Vítkovice steelworks or the Daimler Motoren company in Wiener Neustadt. At the peak of his career in 1916, the Austrian Minister of War offered him the post of managing director of the largest Austrian munitions plant. He accepted this offer immediately, and from 1916 to 1918, he was the driving force behind broad rationalization measures that were implemented in the Pilsen Škoda factory, which greatly contributed to the further objectification of workers' bodies and the factory's large increase in production.[74]

As the ad quoted above reveals, however, similar rationalization experts did not have to be factory employees. Nevertheless, as Oskar Messter showed in "The Poldi Steelworks," these middle-class reformers in hats, suits, and bow ties could stand behind workers during the whole shift and closely supervise the effectiveness of their work. Especially qualified workers, who before World War I were at the heart of the organized working class, were gradually stripped of the most valuable attributes. While before the war they entered into negotiations with their employers on salaries and other issues in a not entirely equal position, but nevertheless as a relevant partner, the wartime transformation of the world of industrial labor quickly divested them of this status.[75] Instead of being the subjects of collective negotiations, they became the objects of rationalization strategies, upon which they did not have any influence at all. Modern scientific findings, together with the very strict legal regulations, created a framework that categorized workers as machines, for which the conditions and the volume of work were established by objective science and not during negotiations they would have been an integral part of.

Basic labor questions, such as when, how exactly, and in what conditions will the worker work were no longer the subject of negotiations between workers and factory owners or shift foremen, but were issued and controlled by rationalization experts and enforced by laws with extreme, harsh punishments. Together with the massive military draft, the new structure of work relationships was a deep disruption of the prewar working-class collective. Many former leaders disappeared from the factories and were sent to the front, and those who remained had their basic competencies—the free provision of manual labor—taken away by the new organization of factory production.

If we turn our attention to the ways in which working-class culture reacted to these deep changes in the social world of labor, it is necessary to proceed from the same premise as in the case of its reaction to the wartime politics of food. Just as in the case of the wartime diet, even in the case of labor working-class culture did not represent an "authentic" environment that would have to take a stand immediately against the ruling discourse. Rather, working-class culture was jointly created by this discourse. Just like the working-class environment before World War I accepted the metaphor of the human body as a motor in terms of food, it accepted it in terms of labor as well.

The influence of the discourse on manual labor and fatigue on the working-class culture in the first years of the war significantly limited its possibilities to extricate itself from the rationalization framework of the wartime economy. In fact, this framework was in many aspects in agreement with the working class's idea of what manual labor really was. As the major weekly of Czech-speaking metal workers *Kovodělník* (The Metal Worker) instructed its readers in April 1916 in its front page article with a large-lettered headline, simply titled "Work": "Normal work matches human nature. Just like in all other aspects, overdoing work is also something that will sooner or later be punished. Working too long and too strenuously will exact a price; it deprives a person of strength, dulls his soul; these things are known and undeniable. This is why all of the modern workers are calling for an adjustment of working hours, adequate breaks during work. … Experience has shown that shortening the workday increases productivity."[76]

The working-class environment thus replicated the amalgamation of human bodies and machines in questions of work, which likewise determined the thinking of many employers and relevant state locations. Entirely in accordance with the prevailing practice of the Austrian state apparatus, the main solution to the problems of wartime production was in the scientifically substantiated rhythmization of the workday, which would enable the usage of all of the available energy that was at the human body's disposal for working. Modern knowledge thus, according to the working class's opinion, would serve as a judge in the escalating argument between capital and labor on the amount of work and the extent of its valuation.

However, direct experience with the everyday materialization of the rationalization discourse, which subjected thousands of workers to desperate experiences of misrecognition and humiliation, began especially during the last war years to disrupt their ideas on the nature of the work they were doing. The loss of the free disposal of their own bodies gradually assumed a central position in the working-class discourse on work and became one of the important starting points for the working class's increasing discontent. In December 1917, the Union of Metal Workers, the largest union organization of Czech speaking workers, stated at its congress that the prevailing situation is "basically the requisition of work. … The worker is deprived of the only economic value of manpower and a requisition is done of his effectiveness without negotiating with him. … The requisition of the worker's manpower is an economic transfer that the government hasn't understood. Workers have no other means of defense than the strike."[77]

Similarly, the union of Austrian miners bitterly noted that: "Wartime laws (decrees), the most inconsiderate execution of other current laws, militarization and the disenfranchisement of all miners that follows from it, decided that the basic rights between the worker and the coal entrepreneur should land in favor of the entrepreneurs, making the miners pay the ultimate sacrifice to the war while the mine owners significantly profited."[78]

The full requisition of the worker's body and his manpower in favor of wartime production regularly appeared as the main topic in many workers' assemblies, union negotiations and the daily reporting on the war; the workers' daily press noted every attempt to break free from the strict disciplinary framework. The main daily of the Pilsen Social Democratic Party, *Nová Doba*, published, for example, an extensive report in January 1915 on the trial of blacksmith Oldřich Pacovský. He had worked until the end of 1914 at the Pilsen Škoda factory, but due to his workplace's distance from his home and his worsening health, he found a new job by January 1, 1915. The Škoda factory, however, refused to release him and give him his work ledger. There was an argument over the interpretation of the relevant law as to whether employees of militarized factories who worked there before the war had the right to resign. The court's verdict in Pilsen on January 13, 1915 unequivocally stated that the law on wartime operations applied to all factory workers with no exceptions and, therefore, that Pacovský must remain at his job for as long as the military administration decided. According to the court, he did not have the right to change employers.[79]

The working-class environment devoted significant publicity to similar arguments that always ended up with the same negative results for workers. Workers' newspapers published legal analyses of the relevant laws with detailed explanations of what they meant for the workers' position in the sphere of wartime production. The loss of the freedom to dispose of one's own body was thus ever present in working-class culture.

This loss had a direct influence on working-class organization, which was a crucial pillar of prewar working-class culture. The outbreak of war in the summer of 1914 represented a deep fracture in the political and social life in the Austro-Hungarian Empire. Parliament was suspended, prewar political parties were cut off from the majority of their public influence and many prewar associations were either banned outright or gradually paralyzed by the drafting of their members. Despite this radical impact on the basic institutions of working-class culture, however, some workers' organizations stayed active during the war years and continued to provide many workers with a basic platform for the cultivation of mutual interests, the creation of social capital and the general maintenance of the institutional base of working-class culture. When the direct political representation of the majority of workers by the Social Democratic Party or some of the other parties that were aspiring to articulate workers' interests played a passive, conformist role vis-a-vis the state and the inner life of the political parties was to a great extent paralyzed,[80] certain labor unions took over the task of the central points of labor organization.[81] They provided a significantly weakened but nonetheless still existing base for working-class culture, enabling it to form a new collective working-class identity and formulate collective demands with regard to the sphere of industrial labor.

From the big number of pre-war union associations, the ones that represented either unions that were not as affected by the draft, or highly qualified professions, which were the core of the "labor aristocracy," characterized by a relatively high salary level and intensive collective organization, played the biggest role.[82] A typical example were the miners' unions, who were not heavily drafted, were never affected by unemployment during the war and whose salary levels allowed them to keep their organization afloat financially, albeit in much more modest conditions than before the war.[83] Such unions were then able to maintain a basic level of institutional grounding of working-class culture, manifesting itself in meetings or the publication of their own print media. Thus, for example, the Union of Austrian Miners, which united Austrian miners from various regions and of various languages, published four periodicals during the war—in Czech, Polish, German, and Slovenian.[84]

When large portions of the Austrian monarchy's industry were under special legislative control, which draconically punished any signs of collective action, the unions did not have to devote themselves to organizing strikes and could turn their attention to the mitigation of the growing salary inequalities that for most workers meant a steep descent down the social ladder and quite often an existential threat as well.[85] The various material support that individual union offices paid out or tried by various means to have them paid, was, next to publishing their own print media or organizing meetings, crucial for the new constitution of a workers' collective.

When a significant number of workers were drafted into the army and when their colleagues were devalued by the prevailing discourse to the position of factory equipment, participation in the financial flows inside the workers' collective was a significant component of workers' group identity. All of the contributors that relegated part of their salary to various union organizations were publicly named in the workers' press and earned a certain prestige among their peers.[86] A contribution sent to the union cash register, or received from the cash register, demonstrated membership in a workers' collective and constituted one of the few reminiscences of their pre-war collective potential to negotiate as an autonomous subject. The ways in which the main unions tried to gain and redistribute financial resources show how slowly they, as well as the whole working class, were able to adapt to the completely new framework of the wartime economy.

Just like before the war, when individual union offices ran out of their own resources in the form of interest-bearing deposits or continuously paid membership fees, they turned to the industrialists' organizations. In this way, they basically tried to replicate the prewar economic arrangement, based on an uneven, albeit in many aspects partner dialog of labor and capital. In March 1915, for example, the union of metal workers made an official appeal to the association of entrepreneurs, in which it called for cooperation in providing financial bonuses to some qualified working-class professions.[87] A year later, the same union submitted a request for a discussion on wartime support for war returnees and their families.[88] In the third year of the war, union leaders' horizons were still defined by the experience of prewar organized capitalism,[89] where giant industrial cartels operated on the one hand, and on the other individual union organizations perceived themselves as another type of cartel, which provided employers with a workforce.[90]

While the various ways of gaining and subsequently redistributing material resources was the central factor in the inward strengthening of the working class, i.e., solidifying its internal cohesion, the central element in its outward definition was the wartime discourse of injustice. It partially sprung from the main union organizations' inability to understand the changed context of the Austrian economy, and some of its main components therefore still retained their prewar form. Throughout the war, the main point was thus the emphasis on salary inequality, which was the most flagrant example of the devaluation of workers' social position.

The environment of the strictly controlled public space of the wartime monarchy, however, significantly limited the possibilities of articulating this workers' discourse. Wartime censorship meticulously controlled all workers' printed matter, and the state bodies were usually present at any workers' meeting.[91] Nevertheless, especially on the pages of workers' periodicals, many more or less hidden referrals to the rising salary inequality within the Austrian economy were repeatedly printed.

In this context, the journalism genre of business and financial reporting was seen in a new light. Before the war, short news reports on the economic conditions of the enterprises were generally published deep within the respective newspapers and did not attract much attention. If we look at the pages of workers' newspapers during the war, however, we can clearly see an increase in their significance from 1915. The publication of the annual reports of the large Austrian factories, the transcripts from the meetings of their board of directors, or their reports on the progress of stock dividends was used to articulate the discourse of income inequality, which at the same time identified the central enemy of the working class in the field of industrial labor. It was a strategy that could be harshly prosecuted. Reports on the financial affairs of joint-stock companies had to be made public, if only to provide information to investors. However, their simple presence in the context of workers' print media was generally enough to articulate the obvious injustices.

The figures of the rising profits of the Austrian munitions plants, mines or machine works were a departure for workers' newspapers, which otherwise almost solely featured reports on the curtailing of food rations, the trials of specific workers, or the straggling salaries in individual branches. Thus, for example, the central paper of the Union of Metal Workers, the weekly *Kovodělník*, published the following passage on February 25, 1915, on its front page: "Prague Ironworks announces that it recently began to work on another tall furnace and it has hired several hundred new workers; the Florisdorf locomotive factory informs the public that it will pay out a 56k dividend per share this year. … Austrian Alpine Company announces that December exceeded expectations and positive correction will continue, so a higher dividend than originally promised can be expected. These are all signs of improvement."[92] Five weeks later, readers of the same newspaper could read about the balance sheet of the Pilsen Škoda factory, which in 1914 showed a profit of 6.422 million crowns and could thus afford to pay out a dividend of 28 crowns per share.[93]

Pointing out the unequal division of the wartime industry profits became the main way to demonstrate the injustice of the wartime economic system. Despite strict censorship, this discourse became gradually radicalized, so that in 1916 the workers' newspapers often did not content themselves with the mere printing of financial statements, but the editors directly interpreted them for their readers. The anonymously written article "Profitability of Ironworks," which appeared on the front page of *Kovodělník* in March 1916, contextualized the division of wartime profits thusly:

> Two of the largest ironworks have just published their yearly financial statements for 1915. They reveal how profitable these enterprises are. How during the war, dreaded from all sides, the situation could be used for enormous personal profit. These are completely unheard of things during a time when large segments of the population are searching for ersatz food to replace the missing normal diet. … We repeatedly hear that there is no milk for children—our next generation—and fears of an insufficient diet aren't petty. Meanwhile, large ironworks are amassing gigantic profits during the war—and especially from it. … The news from the Alpine and Austrian Mining companies inform us of it. The net profit of the Alpine Company, which in 1914 dropped to 8,8 mil. Crowns, rose in 1915 to 19,385 million. … An allowance that all of Austria, including the military administration, paid to the lords of the iron cartel during a time of pervasive suffering. It goes without saying that the board of directors did not come up short in this business. Its bonuses amounted to 1,58 mil. crowns. This is more than the company spent on insurance for all of its workers.[94]

In May of the same year, a headline announced the arrival of "A World of Profits": "One after another, accounting balances are being published and all of them tell of extraordinary net profits … the Pilsen Škoda factory reveals a gross profit of almost 25 mil. crowns for 1915, 9,5 million higher than in 1914 … a net profit of almost ten million Crowns,"[95] stated the author of the front page article in the paper, where the rest of the issue was devoted only to workers' hunger in various parts of the monarchy. At the beginning of the next year, the same newspaper published comprehensive statistics of the development of the mining, iron, and engineering industries in Austria from the year 1916, according to which every branch grew exponentially.[96] In March 1917, the mass resignation of the board of directors of the Pilsen Škoda factory attracted a lot of attention, for each of its members received a lifelong annuity of between 40,000 and 100,000 crowns.[97] Similar cases further strengthened the articulation of the working-class discourse of injustice, which at the same time clearly identified the main enemy of the working class. The state, which drew up the restrictive framework of the wartime economy, was not seen as the enemy. The individual employers who implemented specific practices within it were.

This discourse of salary injustice, which the key unions offices succeeded in establishing as one of the basic defining criteria of the enemy of the wartime working class, was also the means of maintaining an organized class in the field of labor. Similarly to the case of food, the complex reorganization of labor in the wartime economy had a calamitous effect on the working-class organization. Despite the fact that the majority of unions pushed hard to stay afloat, many of them suffered deeply during the first years of the war. Although some unions managed to continue their activities on a basic level, except for a few exceptions such as the miners' union, they were not spared the personnel reduction caused by their members' disappearance at the front. Ordinary members were not the only ones to leave, however; many functionaries of all types also left. For many, being drafted into the war meant an absolute disintegration of their values and therefore a definitive end to their union activities. The result was an endless series of open obligations, which noticeably weakened the strength of their individual union associations.

The broader board of the Union of Czech Metalworkers, for example, made a statement at its meeting in December 1914 about the union being threatened by a "shortage of honest functionaries, especially cashiers, who did not enter the collected fees into the accounts before they left for the war. Our accounting is making a note of this, in order to remind and collect fees of those in question upon their return. … For example, when the cashier Pulec from Boskovice fell in the war, 29K remained unsettled. Due to dishonest cashiers, the towns of Jičín, Julianov and Bystrc have a mess in their cash registers, which must be rectified."[98] The Union of Railway Employees also had to deal with similar problems, as the misappropriation of union funds reached astronomical heights in 1915. Financial deficits of several thousands crowns were not an exception in the accounting of many local organizations. But, as the culprits were away at the front, it was not possible to enforce it.[99] In June of 1915, the control commission of the metalworkers' union officially requested "that comrade Kacafírek return the amounts overpaid to cover his travels and … submit an explanation of how he could make presentations at two meetings on the same day and at the same hour."[100] Kacafírek submitted his answer to this request in April 1917, when he sent the union a "sincere greeting from war captivity," which finally convinced the board of the necessity to definitively write off the missing sum.[101]

Cashiers Pulec and Kacafírek were only two of the many regional functionaries, whose debt the unions had to forgive during the war years. The financial management of the Union of Czech Metalworkers, for instance, quickly developed a deficit of over 10,000 crowns due to the war mobilization during the second half of 1914, but the board of directors was aware that "it is necessary to be prepared for even worse times and we must all be on guard."[102] The western Bohemian unions were even forced to significantly limit the publication of their newspapers due to the difficult material situation. While in 1914 there were thirty-nine union periodicals in western Bohemia, a year later, there were only twenty-eight, most of them in minimal circulation.[103]

The second basic problem that threatened the very existence of the wartime working class was, next to the noticeable personnel and material reduction, the reorganization of the labor world, which stemmed from the discourse of maximum rationalization. The loss of authority over their own bodies, leading to the devaluation of workers to a position of one of the many factory tools, drastically reduced the future prospects of many workers, a situation not dissimilar to the one that workers experienced when departing to the front. Many of them completely resigned from any active engagement and several local organizations started to fall apart in front of their leaders' eyes. In August 1915, the secretary of the union of metalworkers Antonín Hampl complained that "employers currently have great power based on various decrees, which they use to take advantage of the working class, which, bitter, asks the organization to intervene. Due to the current conditions in some cases this is hard to do. … [W]orkers think that what the organization does is futile and worthless."[104] What is interesting in this context is that Hampl was not only referring to the thoroughly restrictive and disciplinary conditions that workers worked in, but also to the fact that the maximum rationalization of production, together with the military draft, basically eliminated unemployment. Already in the fall of 1915, the number of job vacancies in Prague was higher than the number of registered unemployed persons.[105] This noticeably undercut one of the basic legitimizing arguments for the unions' very existence, which consisted of insuring workers against the loss of employment.[106]

The success of the rationalization of wartime production shook the ground under the working-class culture's feet. The precise rhythm of the working day, divided into long shifts interspersed with regularly scheduled breaks, the clearly established time for work and the time for rest left little room for any other activity than regenerating one's physical strength for further work. Many regional union organizations ceased their activities not because they were paralyzed by the military draft or state persecution, but simply because none of their members had the time or energy to take care of everyday administration.[107] In several regions during the first years of the war, individual union members even resorted to making the rounds of all workers' apartments, where they were literally begging people to enter into a union.[108] "You can't imagine how the life of the organization flagged. … A sad sign of the maturity of organized Czech and Moravian Soc. Democrats,"[109] complained one of the leaders of the regional unions in a letter to his friend at the front.

From 1916 onward, the working class reacted to this development with growing anger at the collapse of the prewar partnership system in collective negotiations and its replacement with an arrangement that only benefited employers. Pressured by constantly advancing rationalization, working-class culture began to free itself from the discourse of work that merely emphasized its mechanical-physical components. A wave of rationalization measures affecting many Czech enterprises at the end of 1916 aroused unusually sharp disapproval among the working class: "We don't doubt in the least that the entrepreneurs will accept it [the rationalization of enterprises] with satisfaction. It means nothing to them but the rising of profits. … Military interests will be pretended, and 'business sciences' will be used, from a lack of other tasks, all for the purpose of loading the 'economic utilization' onto the working class' back."[110] With this disapproval, the collective identity of the working class was re-strengthened through identifying the main enemy, who was not primarily the authoritative state or the Austrian military dictatorship, but the employers.

The rationalization of wartime industry gradually caused the total delegitimization of the prewar liberal order within a considerable part of working-class culture. Although harsh criticism of political and economic liberalism was present in the workers' socialist movement from its very beginning, the war and the accompanying radical reorganization of the world of industrial labor provided opportunities to imagine completely new horizons of the future. The original complaints over the violent requisition of work that went against the prewar ideal of equality and freedom of the contractual parties was slowly replaced by complex criticism of the whole prewar social order. As an anonymous working-class editor noted in January 1917: "People were talking all at once about all kinds of freedoms and the world was drowned in a fever of freedom, but in reality it lost its sure footing. … This elevated freedom turned out to be the freedom of a wild beast, in which the strong and powerful had advantages over the weak."[111]

The traditional liberal understanding of freedom was not the only target;[112] so too was the very institution of private ownership, which constituted the basis of the prewar economic order: "Each owner of the means of production produces without knowing what others produce, of what he most often needs, he produces with willful confidence of private ownership, for private ownership is sacred. That under these conditions some goods are produced in great numbers and others less, that today the markets are full and tomorrow deserted, that this gives rise to local market crises and also a general social crisis, these effects of this anarchy have been known for a long time: but this anarchy is not unwelcome by the big speculators, but a good opportunity for big fish, whose passion is to eat smaller fish.... Therefore they demand: the general public must not interfere in the interests of private ownership, but there must be free competition!"[113]

The everyday experience of maximum rationalization in the wartime economy deeply delegitimized not only the current state, but also the state that preceded it. The ubiquitous experience of the devaluation of work, and in many cases even of one's own dignity, paved the way for thinking about a new future that was qualitatively different from the prewar past. In this context, the year 1917 was a decisive turning point, when the organized working class began to turn away from wishes to return to the old order and instead strove to create a completely new one that would not respect even the most basic postulates of prewar liberalism, such as the strict protection of private ownership.

Radical Marxist concepts of social revolution gained relevance. Broad masses of workers in the Bohemian lands did not usually reach for academic Marxist literature, but rather for popular novels, poems or widely published workers' calendars, and the basic building blocks of Marxist ideology were unknown to them or were grasped only in a very simplified and vulgar way.[114] Wartime experience, however, paved the way for a radical imagination that perceived the reconstruction of the social order not as something utopian, but as something achievable within a few years. The working class in the Bohemian lands thus was not more Marxist at the end of the war than at the beginning. What did change was the horizon of the future, which began to appear as something completely different from anything that had been known before.

Most historiography has attributed this radical change in the Czech working class's horizons to the 1917 Russian Revolution, which was the first one to show the practical feasibility of a communist utopia. The Russian example of a workers' revolution, according to this thesis, was so dominant that it caused a complex transformation of many workers' expectations of postwar renewal.[115] This is certainly true to some extent, yet the rapid spread of the communist project as a plausible conception for the postwar order would not have been possible without the internal, autonomous experience of many Czech workers with the radical transformations of the world of work during the war. The influence of the Russian Revolution thus combined with longer shifts within the Czech working class in the fall of 1917, enabling the Marxist socialist project to spread from the debates of social democratic theorists and functionaries into hundreds of factory halls.[116] The communist utopia began to determine the mental horizons of a considerable part of the Czech organized working class, eventually culminating in the founding of the autonomous Communist party during the first years of the independent Czechoslovak Republic.

The wartime politics of work thus caused a disintegration of the working class to a certain extent, but also fundamentally contributed to its reconstitution in the second half of the war. Part of this newly created working class refused to think in the categories of prewar organized capitalism. Instead, it followed its own horizons, which represented an absolute negation of the prewar liberal order as well as employers, and ultimately capitalism itself.

Notes

1. "Ein braves Pferd stirbt in den Sielen." Otto von Bismarck in a speech before the German parliament in February 1881.
2. I thank "Filmarchiv Austria," and Thomas Ballhausen in particular, for providing me with a copy of the film. I thank Dr. Hannes Leidinger from the University of Vienna for providing me with contacts and letting me know about the film.
3. For more about Oskar Messter, see the catalog of the exhibit: Martin Loiperdinger, ed., *Filmpionier der Kaiserzeit* (Basel, 1994).
4. For a broader analysis of the film and its place in the context of the European cinematography of the time, see Frances Guerin, *Culture of Light: Cinema and Technology in 1920s Germany* (Minneapolis, 2005), 48–73; Kimberly O'Quinn, "The Reason and Magic of Steel: Industrial and Urban Discourses in DIE POLDIHÜTTE," in *A Second Life: German Cinema's First Decades,* ed. Thomas Elsaesser, (Amsterdam, 1996), 192–204.
5. For the development of Poldi Kladno during World War I, see Miroslav Kárný et al., *Sto let Kladenských železáren. Příspěvek k dějinám českého železářství a k dějinám dělnického hnutí na Kladensku v letech 1854–1957* (Prague, 1959), 273–304; Karel Klíma et al., *100 let ocelí Poldi* (Kladno, 1989), 28–46.
6. Rabinbach, *The Human Motor,* 2.
7. See James Van Horn Melton, *The Rise of the Public in Enlightement Europe* (Cambridge and New York, 2001).
8. Lukáš Fasora, Jiří Hanuš and Jiří Malíř, eds., *Sozial-reformatorisches Denken in den böhmischen Ländern 1848–1914* (München, 2010); Werner Drobesch, Die "soziale Frage" der Habsburgermonarchie im zeitgenössischen gesellschaftswissenschaftlicen Diskurs, *Moderní dějiny / Modern History* 1 (2012): 1–12; Rüdiger vom Bruch, "Streiks und Konfliktregelung im Urteil bürgerlicher Sozialreformer 1872–1914," in *Streik. Zur Geschichte des Arbeitskampfes in Deutschland während der Industrialisierung,* eds. Klaus Tenfelde and Heinrich Volkmann (München, 1981), 253–70; Sven Thomas, *Gustav Schmoller und die deutsche Sozialpolitik* (Düsseldorf, 1995).
9. Jan Křen, *Dvě století střední Evropy* (Prague, 2006), 245; Werner Drobesch, "Ideologische Konzepte zur Lösung der sozialen Frage," in *Die Habsburgermonarchie 1848–1918,* ed. Ulrike Harmat, vol. 9, Soziale Strukturen, 1419–1463 (Vienna, 2010); Drobesch, "Die Soziale Frage der Habsburgermonarchie,"

1–12; Kurt Ebert, *Die Anfänge der der modernen Sozialpolitik in Österreich. Die Taafesche Sozialgesetzgebung für die Arbeiter im Rahmen der Gewerbeordnungsreform (1879–1885)* (Vienna, 1975).

10. Wolfgang Maderthaner and Lutz Musner, *Unruly Masses: The Other Side of Fin-de-Siécle Vienna* (New York and London, 2008), 56–57.
11. Frederick W. Taylor, *The Principles of Scientific Management* (New York and London, 1911).
12. François Vatin, "Arbeit und Ermüdung. Entstehung und Scheitern der Psychophysiologie der Arbeit," in *Physiologie und industrielle Gesellschaft. Studien zur Verwissenschaftlichung des Körpers im 19. und 20. Jahrhundert*, eds. Philipp Sarasin and Jakob Tanner (Frankfurt am Main, 1998), 355–56.
13. Rabinbach, *The Human Motor*, 19–44.
14. Rabinbach, "Ermüdung, Energie und der menschliche Motor," 286–87.
15. Karin Johannisson, "Modern Fatigue: A Historical Perspective," in *Stress in Health and Disease*, eds. Bengt B. Arnetz and Rolf Ekman (Weinheim, 2006), 3–19.
16. Angelo Mosso, *La Fatigua* (Milano, 1891).
17. Rabinbach, "Ermüdung, Energie und der menschliche Motor," 288.
18. Rabinbach, *The Human Motor*, 136–37.
19. Wilhelm Weichardt, *Über Ermüdungsstoffe* (Stuttgart, 1910).
20. Rabinbach, "Ermüdung, Energie und der menschliche Motor," 290–91.
21. François Vatin, "Arbeit und Ermüdung. Entstehung und Scheitern der Psychophysiologie der Arbeit," in *Physiologie und industrielle Gesellschaft. Studien zur Verwissenschaftlichung des Körpers im 19. und 20. Jahrhundert*, eds. Philipp Sarasin and Jakob Tanner (Frankfurt am Main, 1998), 354–55.
22. The catalog of the Czech National Library keeps a record of almost all of the seminal works of this research, which the Czech- and German-speaking academic communities in the Bohemian lands received soon after its publication. See, for example, Charles Richet, *L´Homme et l'Intelligence: Fragments de Physiologie et de Psychologie* (Paris, 1887); Josefa Yoteyko, *Intróduction a la methodologie de la psychologie pédagogique* (Genéve, 1909); Ferdinand Lagrange, *La médicacion par l'escercice* (Paris, 1894).
23. Angelo Mosso, *Die Ermüdung* (Leipzig, 1892).
24. Angelo Mosso, *Tělesná výchova mládeže* (Prague, 1901).
25. Helena Janderová, "Psychotechnický ústav," in *Technokracie v českých zemích (1900–1950)*, eds. Jan Janko and Emilie Těšínská (Prague, 1999), 135–43.
26. Nathan Zuntz, *Lehrbuch der Physiologie des Menschen* (Leipzig, 1909); Emil Kraepelin, *Die Arbeitskurve* (Leipzig, 1902); Emil Kraepelin, *O duševní práci* (Prague, 1906).
27. Rabinbach, "Ermüdung, Energie und der menschliche Motor," 291.
28. František Mareš, *Všeobecná fysiologie* (Prague, 1894), IV.
29. František Mareš, *Fysiologie* (Prague, 1906), vol. 1, Všeobecná fysiologie, 154.
30. Cf. Gustav Otruba, "Entstehung und soziale Entwicklung der Arbeiterschaft und der Angestellten bis zum Ersten Weltkrieg," in *Österreichs Sozialstrukturen in historischer Sicht*, ed. Erich Zöllner (Vienna, 1980), 123–54.

31. Stenographische Protokolle des österreichischen Abgeordnetenhauses, 112. Sitzung der X. Session am 11. Februar 1887, 4128.
32. Josef Pazourek, *Taylorova soustava organisace práce* (Královské Vinohrady, 1913), 5. For further evidence, see Stanislav Špaček, *Práce a hospodářství: úvahy o lidské a pracovní ekonomii* (Prague, 1918).
33. Otto Smrček, "Expanze technického myšlení," in *Technokracie v českých zemích (1900–1950),* eds. Jan Janko and Emilie Těšínská (Prague, 1999), 37–56; Emilie Těšínská and Jindřich Schwippel, "Masarykova akademie práce," in *Bohemia Docta. K historickým kořenům vědy v českých zemích,* eds. Alena Míšková, Martin Franc, and Antonín Kostlán (Prague, 2010), 286–31, here especially 307–12.
34. Jindřich Fleischner, *Technická kultura. Sociálně-filosofické a kulturně-politické úvahy o dějinách technické práce* (Prague, 1916), 244.
35. RGBl 1912, 1192–99.
36. On the wartime labor legislation see Raimund Löw, "Die Deutsche Sozialdemokratie in Österreich und die Balkankriege 1912/1913," in *Internationalism in the Labour Movement 1830–1940,* eds. Frits von Holthoon and Marcel van der Linden (Leiden, 1988), 410–39; Plaschka, Hasselsteiner, and Suppan, *Innere Front,* 183–88.
37. Demeter Koropatnicki, *Kommentar zum Kriegsleistungsgesetz: [vom 26. 12. 1912] samt Nebengesetzen; in Verbindung mit dem Gesetzestext, den Erläuterungen des k.u.k. Kriegsministeriums, des Landesverteidigungsministeriums, den Beratungsprotokollen des Reichsrates etc.* (Vienna, 1916), 145–52; Ferdinand Hanusch and Emanuel Adler, eds., *Die Regelung der Arbeitsverhältnisse im Kriege* (Vienna and New Haven, 1927), 31–59.
38. Pavel Potocký, *Dělník na válčených úkonech. Jeho práva a povinnosti* (Prague, 1915), 18–22.
39. RGBl 1914, 819–20.
40. RGBl 1914, 819, paragraph 2.
41. Mesch, *Arbeiterexistenz in der Spätgründerzeit,* 43–44.
42. Hanusch and Adler, *Die Regelung der Arbeitsverhältnisse im Kriege,* 122–28; Plaschka, Hasselsteiner and Suppan, *Innere Front,* 189–97.
43. Šedivý, Češi, české země a velká válka 1914–1918, 218–33. For the broader context of the organization of the wartime economy, see Vlastislav Lacina, "Válečné hospodářství ve střední Evropě a v českých zemích za první a druhé světové války," in *Česká společnost za velkých válek 20. století. Pokus o komparaci,* eds. Jan Gebhart and Ivan Šedivý (Prague, 2003), 45–50.
44. The military director of the Ringhoffer factories' notice to the workers on their subordination to military organs from December 15, 1915, in Marie Stupková, ed., *Sborník dokumentů k vnitřnímu vývoji českých zemí za 1. světové války* (Prague, 1994), vol. 2, year 1915, document no. 68, 152–53.
45. Cf. *Berichte der k.k. Gewerbe-Inspektoren über ihre Amtstätigkeit in den Jahren 1914–1916* (Vienna, 1915–1919).
46. ACCP, Okresní správa politická 1850–1927, Karlín (JAF no. 69), karton 309, sign. 15/0-224, Work regulations of the Odkolek company, paragraph 5.
47. Janáček, *Největší zbrojovka monarchie,* 107.

48. Bohumil Řehák, "Za první války ve Škodovce," in *K dějinám závodů V. I. Lenina,* ed. Václav Jíša and Jiří Kodeš (Pilsen, 1962), 165–73.
49. Nedvěd, *Jak to bylo na českém západě 1914–1918,* 181.
50. Habrman, *Mé vzpomínky z války,* 88–89.
51. Kárný et al., *Sto let Kladenských železáren,* 280.
52. Habrman, *Mé vzpomínky z války,* 87.
53. Janáček, *Největší zbrojovka monarchie,* 379.
54. Nedvěd, *Jak to bylo na českém západě 1914–1918,* 168.
55. Václav Čepelák, ed., *Dějiny Plzně II., Od roku 1788 do roku 1918* (Pilsen, 1967), 226.
56. Jahrbuch der österreichischen Industrie für das Jahr 1914, 546. *Bericht der k.k. Gewerbeinspektoren über Ihre Amtstätigkeit im Jahre 1915* (Vienna, 1916), 365.
57. CMAP, 9. Sborové velitelství (1883–1919), karton 131, sign. 75-4/45
58. Ibid., karton 96, č. res. 51558.
59. Nedvěd, *Jak to bylo na českém západě 1914–1918,* 184.
60. *Bericht der k.k. Gewerbe-Inspektoren,* CXLIV.
61. Ibid., 281.
62. Ibid., XLII-XLIV.
63. RGBl 1912, 1081–85.
64. Hanusch and Adler, *Die Regelung der Arbeitsverhältnisse im Kriege,* 26–29.
65. *Kovodělník. Orgán svazu dělnictva zaměstnaného výrobou a zpracováním kovů a drahokovů v Rakousku* 36 (October 3, 1914): 2.
66. *Bericht der k.k. Gewerbe-Inspektoren,* XXXIX.
67. Ibid., 425.
68. Ibid., 458.
69. *Prager Tagblatt* 349 (December 17, 1916): 14.
70. On work rationalisation in Central Europe, see Heidrun Homburg, *Rationalisierung und Industriearbeit. Arbeitsmarkt—Managemen—Arbeiterschaft im Siemens-Konzern Berlin 1900–1939* (Berlin, 1991); Christian Kleinschmidt, *Rationalisierung als Unternehmersstrategie. Die Eisen- und Stahlindustrie des Ruhrgebietes zwischen Jahrhundertwende und Weltwirtschaftskrise* (Essen, 1993); Bruce Kaufmann, *Managing the Human Factor: The Early Years of Human Resource Management in American Industry* (Ithaca and London, 2008).
71. Charles S. Maier, "Between Taylorism and Technocracy: European Ideologies and the Vision of Industrial Productivity in the 1920s," *Journal of Contemporary History* 5 (1970): 27–61; Mary Nolan, *Visions of Modernity: American Business and the Modernization of Germany* (New York and Oxford, 1994).
72. Janáček, *Největší zbrojovka monarchie,* 109–10.
73. Gunnar Stollberg, *Die Rationalisierungsdebatte 1908–1933: Freie Gewerkschaften zwischen Mitwirkung und Gegenwehr* (Frankfurt am Main and New York, 1981); Daniel Nelson, *Managers and Workers: Origins of New Factory System in the United States 1880–1920* (Cambridge, Mass., 1977).
74. For broader biographical informatio see Gerhard Niedermayr and Franz Pertlik, "Hans J. Karabacek. Ein später Nachruf," *Mitteilungen der Österreichischen Mineralgesellschaft* 145 (2000): 15–20.

75. Fasora, *Dělník a měšťan*, 261–99.
76. *Kovodělník. Orgán svazu dělnictva zaměstnaného výrobou a zpracováním kovů a drahokovů v Rakousku* 14 (April 6, 1916): 53.
77. Labor Union Archives, Prague (Henceforth: LUAP), fond Staré odborové spolky, sign. 4592/325, Transcript of negotiations at the Reichs conference of the Union of Czech Metalworkers in Austria on December 31, 1917.
78. *Union of Austrian Miners, activities report for the years 1912, 1913, 1914, 1915, 1916 a 1917* (Trnovany and Teplice, 1918), 2.
79. *Nová Doba* (January 13, 1915): 3.
80. For the internal tension that Social Democracy's passivity provoked in its own membership, cf. the latest Lukáš Fasora, "Generační revolta v socialistickém táboře v letech 1900–1920," *Český časopis historický* 2 (2012): 288–317.
81. On labor unionism in the Bohemian lands, see František Srb et al., *Nástin dějin Československého odborového hnutí. Od vzniku prvních organizací odborového typu do období nástupu k výstavbě socialismu* (Prague, 1963), 95–100; Jaroslava Pelikánová, Vladimír Dubský et al., *Přehled dějin Československého odborového hnutí* (Prague, 1984), 106–13; Jan Galandauer, *Bohumír Šmeral 1914–1941* (Prague, 1986), 28–49; Zdeněk Kárník, *Habsburk, Masaryk či Šmeral. Socialisté na rozcestí* (Prague, 1996), 64–77.
82. Machačová and Matějček, *Nástin sociálního vývoje českých zemí 1780–1914*, 186–89; Vlastimila Křepeláková, *Struktura a sociální postavení dělnické třídy v Čechách 1906–1914* (Prague, 1974), 53–81.
83. LUAP, fond Staré odborové spolky, sign. 3641/274, Report from the meeting of the union of Czechoslovak miners based in Most on July 16, 1916.
84. *Unie horníků rakouských. Zpráva o činnosti za rok 1912, 1913, 1914, 1915, 1916 a 1917*, Teplice-Trnovany 1918, 10.
85. *Právo lidu* 65 (March 5, 1916): 4.
86. See, for example, *Právo lidu* (October 10, 1915): 4.
87. LUAP, fond Staré odborové spolky, sign. 4384/317, Zápisy z jednání představenstva svazu českých kovodělníků, zápis ze (March 17, 1915).
88. LUAP, fond Staré odborové spolky, sign. 4384/317, Zápisy z jednání představenstva svazu českých kovodělníků, zápis z (March 26, 1916).
89. Křen, *Dvě století střední Evropy*, 249. For the European context cf. Marcel van der Linden and Jürgen Rojahn, ed., *The Formation of Labour Movements 1870–1914. An International Perspective* (Leiden, 1990), vols. 1–2.
90. Harm G. Schröter, "Kartelierung und Dekartalierung 1890–1990," *Vierteljahrsschrift für Sozial- und Wirtschaftsgeschichte* 81 (1994): 457–93; Hans Pohl, ed., *Kartelle und Kartellgesetzgebung in Praxis und Rechtssprechung vom 19. Jahrhundert bis zur Gegenwart* (Stuttgart, 1985); Geoffrey Jones, ed., *Coalitions and Collaboration in International Business* (Adlershot, 1993); Elizabeth Domansky, "The Rationalization of Class Struggle: Strikes and Strike Strategy of the German Metalworkers' Union, 1892–1922," in *Strikes, Wars, and Revolutions in International Perspective: Strike Waves in the Late Nineteenth and Early Twentieth Century*, eds. Leopold D. Haimson and Charles Tilly (Cambridge and New York, 2002), 321–55.

91. Galandauer, *Bohumír Šmeral 1914–1941*, 28–33.
92. *Kovodělník. Orgán svazu dělnictva zaměstnaného výrobou a zpracováním kovů a drahokovů v Rakousku* 8 (February 25, 1915): 29.
93. *Kovodělník. Orgán svazu dělnictva zaměstnaného výrobou a zpracováním kovů a drahokovů v Rakousku* 14 (April 8, 1915): 59.
94. *Kovodělník. Orgán svazu dělnictva zaměstnaného výrobou a zpracováním kovů a drahokovů v Rakousku* 11 (March 16, 1916): 41. For the context of the wartime development of the Alpine Mountain company, see Otto Hwaletz, "Österreichisch-Alpine Montangesellschaft bis in die 1930er Jahre," in *Business History: Wissenschaftliche Entwicklungstrends und Studien aus Zentraleuropa*, eds. Alice Teichová, Herbert Matis, and Andreas Resch (Vienna, 1999), 229–46.
95. *Kovodělník. Orgán svazu dělnictva zaměstnaného výrobou a zpracováním kovů a drahokovů v Rakousku* 19 (May 11, 1916): 73.
96. *Kovodělník. Orgán svazu dělnictva zaměstnaného výrobou a zpracováním kovů a drahokovů v Rakousku* 3 (January 18, 1917): 11.
97. Janáček, *Největší zbrojovka monarchie*, 366.
98. LUAP, Staré odborové spolky, sign. 4384/317, Transcripts of the meetings of the Union of Czech Metalworkers for the years 1914–1916, transcript from December 6, 1914.
99. LUAP, Staré odborové spolky, sign. 3010/229, Ledger of meetings of the union of rail attendants, transcript from April 24, 1915.
100. LUAP, Staré odborové spolky, sign. 3010/229, Ledger of meetings of the union of rail attendants, transcript from June 27, 1915.
101. LUAP, Staré odborové spolky, sign. 3011/229, Ledger of meetings of the union of rail attendants, transcripts from April 22, 1917.
102. LUAP, Staré odborové spolky, sign. 4384/317, Transcripts of meetings of the Union of Czech Metalworkers for the years 1914–1916, transcript from December 6, 1914.
103. Nedvěd, *Jak to bylo na českém západě 1914–1918*, 263.
104. LUAP, Staré odborové spolky, sign. 4385/317, Transcripts of meetings of the board of director of the Union of Czech Metalworkers fro the years 1916–1917, transcript from August 15, 1915.
105. *Právo lidu* 330 (November 30, 1915): 4.
106. LUAP, Staré odborové spolky, sign. 4385/317, Transcripts of meetings of the board of directors of the Union of Czech Metalworkers for the years 1916–1917, transcript from August 15, 1915.
107. LUAP, Staré odborové spolky, sign. 4385/317, Transcripts of meeting of the board of directors of the Union of Czech Metalworkers, Transcript of the meeting of the wider board of directors from November 19, 1916.
108. Report of the Prague co-ownership from July 8, 1916, in Rudolf Neck, ed., *Arbeiterschaft und Staat im Ersten Weltkrieg I. Der Staat* (Vienna, 1964), document no. 63, 83.
109. ACCP, Paměti Vojtěcha Bergera, Z Války, Kniha II., Odborová, od 1. ledna 1914 do 31. prosince 1926, Transcript from September 7, 1915, 43.
110. *Kovodělník. Orgán svazu dělnictva zaměstnaného výrobou a zpracováním kovů a drahokovů v Rakousku* 1 (January 4, 1917): 2.

111. *Kovodělník. Orgán svazu dělnictva zaměstnaného výrobou a zpracováním kovů a drahokovů v Rakousku* 3 (January 18, 1917): 9.
112. See the classic work Isaiah Berlin, *Čtyři eseje o svobodě* (Prague, 1999).
113. *Kovodělník. Orgán svazu dělnictva zaměstnaného výrobou a zpracováním kovů a drahokovů v Rakousku* 13 (April 1, 1915): 53–54.
114. For a very limited reception of Marxist thought within prewar working-class culture, cf. the classic work Dieter Langewiesche, *Zur Freizeit des Arbeiters. Bildungsbestrebungen und Freizeitgestaltung österreichischer Arbeiter im Kaiserreich und der Ersten Republik* (Stuttgart, 1979), 174–222; Peter Kuleman, *Am Beispiel des Austromarxismus. Sozialdemokratische Arbeiterbewegung in Österreich von Hainfeld bis zur Dollfuß-Diktatur* (Hamburg, 1979), 21–31; Jakub Beneš, "Socialist Popular Literature and the Czech-German Split in Austrian Social Democracy 1890–1914," *Slavic Review* 2 (2013): 327–51.
115. Galandauer, *Bohumír Šmeral 1914–1941*, 145–62; Jan Galandauer, *Ohlas Velké říjnové socialistické revoluce v české společnosti* (Prague, 1977).
116. See the latest Juraj Benko, *Boľševizmus medzi východom a západom (1900–1920)* (Bratislava, 2012), 83–90.

Chapter 3

Rationed Manliness: The Politics of Gender

"God, I want to be a man,
but I don't want to die with a rifle in my hand."
—Bad Religion, "Heaven is Falling,"[1]

A World Without Men

Friday, September 11 to Friday, September 17, 1914 was a very busy time for the *Peklo* (Hell) workers' community house in Pilsen. The house, which had become the center of workers' social and cultural life before World War I,[2] screened the newest blockbuster from the studio of one of the most famous German directors, Max Mack, entitled *Die Welt ohne Männer* (A World Without Men), every evening.[3] The leading role in the film was played by the British actress Madge Lessing, one the biggest theater stars in prewar London, Paris, Berlin, and Broadway.

The film itself had great potential for commercial success and mass appeal. It was not based on a famous book or popular motif, but on a script written especially for the film by Alexander Engel and Julius Horst. As a film based on an original screenplay, *Die Welt ohne Männer* offered the most modern example of what the cinema of the time could offer. Films based on original screenplays were quite popular before the war, and any director who aspired to international recognition made them.[4]

Die Welt ohne Männer was a comedy about a woman who refuses to be in any kind of relationship with a man. For her, marriage was an institution that merely confirmed the pointless domination of men, and men were the embodiment of cruelty. However, during the film, the heroine meets an attractive suitor with whom she falls in love. Their burgeoning relationship forms the basis of many comedic situations, during which the heroine's ideas about men as beasts are confronted with the real figure of the handsome and practically perfect boyfriend.

The film parodied aspects of the women's emancipation movement, which had become one of the liveliest social topics in the Bohemian Lands before World War I.[5] Women's emancipation was caricatured as a list of demands that were based on unreasonable fantasies; women's demands for certain rights were portrayed as a needless endeavor since men were so charming and caring.

The film became very popular among Pilsen workers. Showings sold out fast and local newspapers competed with one another to print invitations to a relaxing comedy with humorous situations, in which "the heart of this [emancipated] woman burns with a love that develops in a comical way."[6] The movie evidently conformed to the tastes of *Peklo*'s audiences. The humorous parody of the current women's emancipation efforts fit in with the male workers' worldview, to whom the ads were addressed. But the *Peklo* workers' house was not a space reserved only for male workers. Almost every week, it held events for the workers' wives, either in their roles as housewives, or as industrial workers. Such events, however, were meant exclusively for women, which is also how they were advertised in the special sections of workers' newspapers. But *Die Welt ohne Männer* was a fun film for men and was thus advertised in the respective sections of all the local working-class dailies. Workers were invited to "come see and laugh"[7] at the inept heroine, whose crazy ideas about the rotten character of men led her into various uncomfortable situations, which in the end were supposed to reassure viewers that, in a world without men, "most women would surely miss them."[8]

The publicity that the film received among Pilsen workers illustrates the broader context of gender relations at the beginning of World War I. Laughing at aspects of the women's emancipation movement was one of the possible defensive reactions male collectives adopted when confronted with the changing gender order. The instability of men's traditional positions as bearers of rationality, familial authority, and active subjects of public life resonated in the years before World War I in almost every social environment. The war experience then quickly created a new context, in which the gradual, rather slow prewar transformation of the gender order became a veritable whirlwind of changes affecting numerous aspects of everyday life.[9]

As we have seen in the previous two chapters, the wartime restructuring of Austrian society deeply affected the world of food consumption, as well as that of industrial labor. The strict rationalization of food and working hours had a daily impact on the very foundations of workers' self-identification. Next to the spheres of food and work, the politics of gender constituted a third front in which the organized working class was forced to fight for its very existence. Just as workers' self-identification was challenged by the supply crisis and the wartime reorganization of the economy, it was also undercut by the far-reaching changes in gender relationships.

The previous chapters focused on the transformations of the working class in the area of consumption and work, i.e., on two prominent stages that shaped the working class in the pre-war era. The politics of food and work were two closely connected spheres that emerged in large part from a shared scientific discourse on optimized consumption and production, structured around the amalgamation of mechanical machines and working bodies. This discourse cannot be ignored in the third significant sphere of gender either. Despite that working-class masculinity and femininity and their concomitant interrelationships were not constructed only within the discourse of modern positivist science; they also took shape, in many regards, under the influence of several modern scientific fields. An analysis of the politics of gender thus cannot be completely separated from the politics of food or the politics of work. In fact, masculinity and femininity were often constructed in close interaction with the consumption of food and the practices of industrial work. The politics of gender were entwined with the world of consumption as well as that of labor long before the outbreak of World War I. However, the wartime era ushered in radical and profound changes that significantly transformed the organized working class. Next to decreasing food rations and increasing rations of industrial labor, a large portion of the heretofore androcentric working class had to adapt to decreasing rations of their own positive male self-identification.

Publicly Organized Motherhood and the New Familial Authority

The concept of the body as an energy-transforming motor treated food as an entry source and manual labor as the final product. At a time when an increasing percentage of industrial production became rationalized and mechanized, the human body was seen as one of the factors in economic growth equation. The combination of effective work, quality technical equipment, and overall work organization played a key role in the success of a given enterprise. But even before the war, gender relationships complicated the discourse that saw all human bodies through the universalizing lens of economic logic.

Middle-class scientists and social reformers had been attempting to overcome the conflict between maximum efficiency and gender difference since the formulation of the "social question" in response to specific phenomena connected to European industrialization.[10] The middle classes in particular tried to integrate the urban poor and the working class into their own set of values. At the turn of the twentieth century, educational campaigns to introduce modern science to the lower urban classes and integrate them into the rationalized and rationalizing middle-class world intensified. These campaigns took place to a great extent along established gender divides that relegated reproductive work in the household to women not only in the middle class but also in the working class, where, for economic reasons, the ideal of separate spheres could never be fully implemented.[11] Between the years 1914 and 1918, campaigns for the right diet therefore almost always focused on working-class women.

Envisioning women as objects of scientific advice (usually formulated in the middle-class male environment of various scientific fields) had an ambivalent effect. On the one hand, it confirmed the primacy of modern science and the middle class that produced it by discovering objective norms for organizing everyday life. On the other hand, however, keeping male workers ignorant of rationalized nutrition significantly increased the importance of working-class women, who had access to knowledge that remained hidden from their male counterparts.[12] During the war, this knowledge of optimal nutrition became paramount for working-class women who had to make sure that available food was purposefully utilized. As the discourse on food rationalization spread, so did women's opportunities to express themselves publicly. As dietary reformers, they could participate in public discussions that were previously only open to men.

A completely new arena of public engagement was opened up for women, partially legitimized by their traditional roles as mothers and housewives. However, the very act of taking on these roles removed them from the private spaces of the household and turned some of them into public figures. Maternal love and the responsibility for the household, as well as the childrearing it entailed, effectively became a job, redefining the roles of many women. Especially for unwed or widowed women, it opened up new possibilities of public engagement.[13]

In the Bohemian lands, the embodiment of this "publicly organized motherhood"[14] was Anuše Kejřová. A cooking and home economics teacher who became a legend in Czechoslovak women's education and home management between the wars, she represented this new type of publicly engaged housewife in the decades before World War I and during the war itself. Unlike preceding generations of housewives, such as Magdalena D. Rettigová, she did not limit herself only to writing cookbooks, and her activities broadly affected the Czech public space during the war. In 1910, she obtained a license to open her own culinary school, which she led during the war; she taught cooking and home management classes at a girls' school in the eastern Bohemian city of Hradec Králové (Königgrätz); and she regularly organized traveling home management classes all over the Bohemian lands.[15] Unlike Magdalena D. Rettigová, her recipes emphasized frugality as the main principle,[16] while her practical cooking courses for poor and working-class housewives were basically a model laboratory in which the middle-class discourse of rationalization was introduced to the working class. Every week in her classes, Kejřová taught her students rational use of available sources, cost-saving meal preparation, as well as modest consumption.[17] These traveling cooking classes flourished enormously during the war years and occurred at least once a month in every larger Czech town.[18]

The significance of similar educational events for working-class women's changing status within their own families was enormous.[19] The knowledge disseminated through these channels was seen at the time as the embodiment of objective, modern, and scientific findings, and women became agents, who were in many respects not only equal to men, but sometimes even surpassed them. As the Prague doctor R. A. Suk stated: "An aware working-class woman will first and foremost take care of the proper healthy lifestyle of her family, for the woman has more opportunities to change the family's situation than the man."[20]

Working-class women thus became the recipients of the newest discoveries of nutritional science—a significant factor in the rationalization of food consumption, as well as in the war effort itself. Working-class families experienced new situations in which the main authority for solving everyday issues was the working-class housewife equipped with the objective knowledge to organize many aspects of everyday life. Male workers were stripped of their freedom to decide what, and when, to eat. It was now their partners who determined the manner and time of meals. They alone knew how to ensure that these everyday activities would be carried out with the greatest possible effect.

The monthly journal *Zájmy žen* (Women's Interests), for instance, instructed its readers to apply the newest findings on optimum food consumption at home:

> Above all remember that we should never trouble the stomach with a new meal before the previous one has been completely digested. … Secondly remember that it is necessary for meals to be eaten with the utmost regularity. A healthy, adult person should eat three times a day. … We should eat slowly, chew thoroughly, coat every bite with saliva and calmly swallow:

> a bite of food should prepare itself for the stomach while still inside the mouth. In the mouth, the bite of food goes through an important mechanical and chemical transformation, which is absolutely necessary for proper digestion. … Only a person who follows these rules can tolerate the current hard times, and will happily get through them to better days.[21]

The former practice of adapting the time and manner of family meals mainly to the male breadwinner yielded to the newer scientific knowledge on human digestion. The bearer of this knowledge, however, was no longer the man, but his partner. "Thanks to mother's diligence, we never went hungry," Robert Matula, a miner from the northern Moravian city of Ostrava, recounted in the interwar period. In one sentence, he captured the transformation of women's roles in working-class domestic culture.[22]

This transformation was, of course, a very sensitive issue for male workers. It caused a visible crack in the existing gender order, in which men were the recipients of scholarly advice for the working class. As the public representatives of working-class households, only men could outwardly represent the family. However, as the working-class press often stated, wartime changes turned this order upside down:

> The revolution of our economic thinking; our economic action is perfected. A person's most private thing, his home, became the crucial point of public interests and government care! The national economy, which, since man began to think, has revolved around the production and sale of goods, concentrates completely on the home, yes, it's becoming a people's home, a national home through production and distribution. Before, everything was arranged in terms of trade, in terms of men; now the focal point is on consumption, which up to now has been the quiet, closed, confidential and almost secret domain of the woman. She ruled over the pantry, closet and cellar, and hers was the important task of dividing the supplies at hand … the woman became much more of a public person than the man.[23]

"Publicly organized motherhood" thus provided domestic housewives with knowledge that frequently disrupted the traditional position of men as the respected bearers of public, as well as private, authority. It became increasingly common for wives to possess authoritative knowledge on how to organize the everyday life of their families. The sphere of food consumption was the first place where men lost their authority—not only within the family, but also in public, where women newly assumed the roles of enlightened agents of necessary changes. Seemingly trivial, it nevertheless contributed to a much broader redefinition of gender relations in the public and labor spheres. From 1914 to 1918, these relations probably went through the fastest and deepest change in all of modern history, resulting in the total destruction of the androcentric character of the Czech working class.

FIGURE 3.1. Vláda žen (The Rule of Women). *Večerník Práva Lidu* 67, 22. 3. 1916: 4.

The Shine and Misery of the Uniform: Gender and the Public Space

"When a heroic wounded or sick soldier arrives at the hospital, the attentive care of the nurse begins,"[24] wrote Robert Marschner in his manual on how to care for Austrian military returnees. In one sentence, he captured the gender order that dominated the public space of the Habsburg Monarchy in the years 1914–1918. Wartime society was symbolically divided into the male world of the front and the female world in the hinterland, which provided indispensable support to the male war effort. If we look at the historiography of World War I, we can see that most of the scholarship basically accepts this separation. Many historians have observed that soldiers had a completely different experience at the front than women in the home front. The front is generally understood as a purely male sphere, while the home front is considered to be an area defined by the absence of men, and thus greatly feminized.[25]

The vast majority of contemporary World War I historiography, as well as the historiography of gender, thus situates the collapse of masculinity, and consequently the unprecedented destruction caused by the war, solely in the sphere of the front, which undermined men's previous identity structured around the peaceful values of liberal progress.[26] To a certain extent, Czech historiography also adopted these conclusions. The most current synthesis of Czech history between the years 1914 and 1918, by Ivan Šedivý, is explicitly divided into a "war front" part and a "home front" part. At the same time, Šedivý attributes the rising social chaos in the Bohemian lands during World War I to precisely this shortage of men and the "male authority" in the family as well as in the public space.[27]

Overall, we can say that in the context of the history of masculinity, historians have thus far focused primarily on experiences at the front, and in most instances conflated man and soldier. As a result, men's experiences on the home front were more or less overlooked. To a certain extent, a large part of the existing historiography has actually fallen under the so-called "wartime experience myth," which was created in the 1920s primarily in Germany and Austria, placing men's sacrifices at the war front before the suffering on the home front.[28] While the historiography of "front masculinity" is thus rather advanced within the context of World

FIGURE 3.2. Men's Front, Women's Home: Austrian Wartime Postcard

War I history writing, the analysis of the redefinition of masculinity in the hinterland lags somewhat behind.

In the historiography of labor and the working class, gender aspects have been similarly long overlooked. Since its emergence, labor history has been characterized by its emphasis on the history of the working class as a unified social formation, whose actions were interpreted through an analysis of working-class material conditions. These were then seen as the basis upon which the subsequent working-class collective action was understood.[29] An emphasis on the material determinants of working-class actions variously dominated the historiography on both sides of the former iron curtain and has not quite disappeared from current scholarship.[30] Attempts to connect the history of the working class with the history of gender thus increased only within the last fifteen years, while the main focus on Western Europe and the United States prevails.[31] The historiography of the working class in central Europe continues to be dominated by a rather social-historical approach, and gender reflexive studies are the exception.[32]

Maureen Healy's previously mentioned work on wartime Vienna is probably the most important recent publication on the Habsburg home front. Healy notes that the notion of a feminized home front in Austria was actually more the product of official propaganda than actual everyday experience. Men who remained at the home front during World War I were in no way a marginal segment of the wartime society. According to Austrian wartime legislation, all men between the ages of 18 and 50 (later changed to 52) could be drafted into military service. However, in 1917, 8,420,000 men out of the overall number of roughly 12 million men capable of military service received their marching orders, i.e., approximately 70 percent. Between 1914 and 1918, more than 3.5 million men between the ages of 18 and 50 did not go through military service. Of the remaining more than 8 million, however, we cannot be sure how many actually fought at the front during the war. Official Austrian statistics from 1917 show that up to that year 4,010,000 men left the army for the following reasons: 780,000 fell at the front; roughly 1,600,000 were prisoners of war; 500,000 were war invalids; roughly 130,000 were past the age limit for military service; 400,000 men were drafted to work in the wartime industry; and 600,000 men left the army for other, unspecified reasons.[33]

Of this large group, we can be sure that only the 780,000 fallen, 1,600,000 prisoners of war, and 500,000 war invalids had direct experience with combat. However, the Austrian law on wartime operations from 1912 allowed a large part of the civil sector to come under military oversight. Many men officially on active military duty thus actually never experienced frontline battles but were drafted to work at the home front for the entire duration of their service. For the majority of men, military service and experience at the front during the years 1914–1918 did not in fact mean the same thing.[34]

Nevertheless, immediately after the outbreak of war, the dominant image in the public space of the Austro-Hungarian monarchy was that of the ideal male recruited soldier, who quickly acquired the position of hegemonic masculinity and replaced all other forms of male self-identification.[35] The model of a man both strong and morally conscientious enough to fulfill his highest duty in battle at the front flooded the public space even before the material consequences of the war were felt on the home front. In this context, the process of military recruitment had a special function—military doctors drew a line between men capable of serving and those who for various reasons could not be recruited.[36] Men who did not pass the military recruitment medical examination quickly began to be called "inferior" (*minderwertig*) and were shamed and humiliated at practically every step.[37]

The process of military recruitment, which separated the men who were able to serve from those who were not, thus established a new definition of hegemonic masculinity and became the general test of full-fledged manhood.[38] The ruling wartime discourse referred to men who were not drafted to the front as people who were unable to adequately sacrifice themselves due to physical reasons or psychological disorders. Within the context of the era's ruling eugenic medical paradigm, such a man was automatically equated to a worthless individual with no social value.[39] Renowned Austrian doctor Julius Tandler wrote in the central Austrian medical journal that "the selection of individuals for war proves advantageous for the transmission of physical weaknesses [among] those left behind, meaning those persons found unfit for military service. Among them one finds persons with sensory irregularities and constitutional anomalies, which are very often inherited. That results in further damage in the next generation."[40]

Any kind of work performed at the home front was therefore drastically devaluated. Compared to frontline battle, all other professions that were masculine before the war now seemed unfit for a real man. The idea that the best men leave to fight at the frontlines while the lesser remain at home defined not only the discourse of eugenic medicine but the environment of the working class as well. A poem that circulated among Czech-speaking workers during the first half of 1915, "Jan Broskovec," by Ladislav Caha, illustrates this:

> He was a handsome lad and his bright gaze observed the world
> The sunny land, the gray clouds that chased storms
> He had a great, strong chest—Jan Broskovec
> He worked in the mine for his masters, he dug, got stronger

Deaf to the world, he knew no shame, he lived his life calmly
Like a bird in a cage—Jan Broskovec
The time came, men in silver collars sent him to war
Gave him blue pants, a shirt and a sword as fallen snow
He was a good dragoon—Jan Broskovec
Then there was a battle, gray regiments crashed into one another
On gray horse the platoon rides, not thinking about impending death
A tragic thing occurred—Jan Broskovec fell[41]

As is evident in these verses, even in the working-class environment the main defining categories of masculinity were structured around recruitment to the front. The strong, handsome men died honorably, while the weak and handicapped stayed home. Almost immediately after the outbreak of the war, the central Czech metalworkers' weekly conflated the recruitment of Czech workers with a valiant, civilizing mission, in which each dignified man had to participate:

> Thousands of our members had to leave their wives, children and work, in order to fight against the lying and cunning enemy in the southeast, and against Russian tsarism that murders both man and culture, that shamelessly dips its hands in the blood of many thousands of people yearning and thirsty for freedom, and that rushes to assist its southeastern ally. We wish with all of our hearts to put a long deserved end to this tsarism.… Culture, built by hundreds of years of struggle, is most alarmingly threatened by Russian tsarism. *To avert this danger is the greatest responsibility of all freedom-loving people fighting for most sacred freedom.*"[42]

As the leader of the Czech Social Democratic party Gustav Habrman later remembered, "[T]housands of beautiful boys of twenty to twenty-four years of age, men who heeded the first call to recruit … Boys and men so strong! A joy to behold. Thin, beautifully strapping, with a twinkle in their eyes.… Boys and men full of energy and strength"[43] set out on this mission in the fall of 1914. According to him, "[T]he nation's core, its muscles, hearts and brains"[44] went off to war, which understandably meant that the opposite remained at home, i.e., men who were thoroughly useless.

This discourse, which equated physical ability and moral integrity with military service at the front, devalued not only men who did not pass the medical examination but also those who were capable in all aspects of military service and were often drafted but for various reasons were left at home during the war. These were firstly workers who lived out World War I away from frontline battles and who were in a paradoxical situation for the entire duration of the war. Since it was not always possible to distinguish through everyday contact in the public space those who had not been drafted from those who stayed at the home front to work in industrial jobs, hardworking workers often faced the dilemma of how to

FIGURE 3.3. "A soldier is the most beautiful man in the whole country." German wartime postcard

remain a man according to the dominant public discourse that automatically identified full-fledged manhood with a soldier making the ultimate sacrifice at the front. Even when various groups sometimes emerged from the anonymous mass of secondary men at the home front who deserved special attention or help, workers were not among them. Austrian instructions for nurses, for instance, established that if it was necessary to provide health care at the home front to someone other than a frontline fighter, absolute priority should be given to "a government clerk or civilian, who is independently gainfully employed."[45] No other social class offered a sufficient enough sacrifice to the total war effort of the monarchy to be allowed to benefit from the health care system, which was primarily designated for army use.

The equation of hegemonic masculinity with a soldier permeated not only through public but also into many private relationships. Many working-class girlfriends demanded photographs of their boys in military uniform for remembrance, and later snapshots of their activity at the front. In September 1914, carpenter Vojtěch Berger's girlfriend Mařka Ctiborová demanded, "that I get my picture taken with my friend Zavadil … in our uniforms,"[46] even though she had enough photographs of him in civilian dress. To have a sweetheart deployed at the front in military uniform quickly became the goal of a young woman whose relationship with a man at the home front would probably fulfill her better emotionally, but who would deem such a male counterpart a lesser man. A photograph of a suitor in military uniform thus confirmed not only male but also female physical fitness, demonstrated through a relationship with a full-fledged partner.

The self-identification of male workers who stayed at home showed its first serious cracks in the fall of 1914, while other changes in the wartime society only deepened them. The image of the man-civilian was falling apart in the public space, and the destructive war dealt serious blows to the entire representation of the imperial-paternalistic order that dominated the Habsburg Monarchy in the decades before World War I. Ernst Kantorowicz, in his world-renowned study of medieval political theology titled "The King's Two Bodies," revealed how, in medieval thought, the sovereign had two different bodies. While the first, physical body was susceptible to the same effects as the bodies of his subjects, meaning that it could get sick, be wounded, and die, the king's second body—the political one—extended beyond his physical form and was the basis of the feudal monarchy's complex social organization. The whole social order was derived from it, and all of the representatives of monarchic power were the representatives of this transcendental royal body.[47]

Although the Habsburg Monarchy at the beginning of the twentieth century had little in common with medieval kingdoms, even here public power symbolically stemmed from the specific person of the ruling emperor. The army, the police, the bureaucracy, and other bodies of state power were officially titled "imperial (and) royal" (kaiserlich und königlich) and received their authority from the person of the sovereign who not only merged with the Austro-Hungarian state as its ruler but also became the embodiment of its fate during the entire second half of the nineteenth century.[48] The organization of Austrian society during the reign of Franz Joseph I was thus often interpreted through the metaphor of the family, at the head of which was a caring but authoritative father.[49] State authority symbolically belonged to the father figure of the emperor, who in many respects embodied Austrian statehood. The very foundations of the state organization were thus already gender coded and the state "grammar of masculinity" was often invisible, but it all the more authoritatively predetermined who, when and where could become the representative of imperial fatherly authority.[50] The fatherly authority of the emperor in the public space of the monarchy was therefore adequately represented exclusively by male figures, who were publicly active people and bearers of public power and social respect. From uniformed officers, policemen, or, in certain circumstances, even significant bureaucrats up to regional school boards or inspectors—all represented the imperial, and therefore male, authority that dominated Austrian streets and offices.

But beginning in September 1914 this symbol of the social order, with men as the main bearers of public authority, began to fall apart. The former order, in which certain professions often connected with high qualifications or public authority were solely male, began to show cracks. On the day of Austria's mobilization, July 31, 1914, Pilsen's *Nová Doba* reported on the forthcoming changes in Prague's municipal transportation: "The Prague board of directors of electrical enterprises will train according to the American model roughly 300 women to drive tramway cars to ensure that there would be substitutes for the male drivers, who where drafted into the war."[51] The Prague transportation authority was likely the first, but in no way the only, municipal transportation agency to undertake this step. In fact, the feminization of controller professions, including the drivers of the municipal electric cars, was obvious in almost all larger Czech cities. Thus, in 1917, the Pilsen transportation authority employed not only many women controllers, but also nine women electric car drivers.

Significantly, uniformed professions that were, as in the case of the controllers, connected with the authority to control tickets and impose fines, or as in the case of the drivers with the responsibility for the whole car, were gradually taken up by women. And they took over not only male po-

FIGURE 3.4. Group photo of the drivers of the Pilsen transportation authority in 1903. Source: Jiří Rieger, Jiří Kohout and Petr Mazný, *110 let Pilsenských městských dopravních podniků* (Pilsen Municipal Archives, 2009), 11.

FIGURE 3.5. Group photo of the women drivers of the Pilsen transportation authority from 1917. Source: Jiří Rieger, Jiří Kohout and Petr Mazný, *110 let Pilsenských městských dopravních podniků* (Pilsen Municipal Archives, 2009), 14.

sitions but also the outer, symbolic signs of masculine authority. While in Pilsen there was a special skirt uniform for women controllers or drivers, some transportation enterprises did not have different women's and men's dress codes and gave even the women employees "pants and jackets for the job, which is usually hindered by women's dress."[52]

Women did not just assume previously male professions that defined the public space at the communal level; they also quickly made inroads into the central institutions of the Austrian state. In September 1915, soldiers on their way to the front in train cars of the Austrian state railway noticed with surprise that "there are women controllers here now."[53] The rising number of women in train and tramway controller uniforms, or in the pants suits of the municipal coach drivers, disrupted the gender order, just like women working as court stenographers or regional and state office clerks.[54] During the year 1916, the first women inspection assistants began working in Austrian trade inspectorates in Lower Austria, Galicia, and Brno.[55] They were first tasked with overseeing enterprises with a majority of female workers, but their activities and numbers increased throughout the last years of the war.[56]

Symbolic male dominance over the public space truly began to recede when women could be found in the very embodiment of the male world—in the imperial army. While at the beginning of the war, a woman in the armies of the countries at war with a position other than health care worker was something positively scandalous, in 1917, over 35,000 women served in Austro-Hungarian "assistance corps" (Hilfskräfte im Felde). Their numbers grew during the war, so that during 1918, the Austrian army numbered almost 50,000 active women soldiers. They did not take part directly in combat and held primarily supportive functions at the home front, such as manning the telephones and telegraphs or as typists. All of these women, however, wore a military uniform, were officially recognized as "drafted,"[57] and could even earn military honors. The Austro-Hungarian army also was not the only one that used women for military purposes during the war. For example, the Russian tsarist army went even further when it incorporated special women's units even in general combat operations and thereby registered the first female losses at the front.[58]

FIGURE 3.6. Decorated woman soldier from 1915. Source: Nová Doba (May 11, 1915) 3.

This entirely unusual engagement was a sensitive matter in the working-class environment. While right before the war, women wearing pants suits occasionally sparked public scandals that would sometimes lead to physical attacks, women in men's clothing, as well as in various uniforms symbolizing public power, became everyday occurrences in the second half of the war.[59] If in January 1915, the working-class press openly made fun of the originally German ideas of putting together women's military units,[60] by 1917, workers at the Austrian home front were forced to accept everyday situations in which women assumed superior positions. Women in male work dress or uniforms were authorized to check fare tickets, impose fines, accept official requests for wartime social support, or decide, in their positions as trade inspectors, which working conditions were acceptable. At the end of the war, workers in Austrian cities had more and more opportunities to meet women even in military uniforms, who, given the hegemonic ideals of the proper man as soldier, cast further doubt on their masculine self-identification. Members of women's army auxiliary units presented a starker contrast. The more they wore their uniforms on the city streets, the more the other signs of former male public authority disappeared. For example, in Pilsen, local policemen stopped wearing an officially required and clearly identifiable uniform in 1915, replacing it with a simple armband.[61] This led to situations in which it was impossible to recognize a policeman who was not in uniform, but women were immediately recognizable in a clearly identifiable military uniform. The close tie between the symbols of public order and authority, on the one hand, and the male gender, on the other, was thus completely disrupted.

Assigning public respect to uniformed soldiers, along with the total collapse of male public authority at the wartime home front, thus completely overturned the ingrained gender order that workers had been used to in the prewar years. Their male self-identification, however, was not only shaken with respect to their status as bearers of public authority and the wartime discourse of the man-soldier as the only variation of full-fledged masculinity. As a matter of fact, workers spent increasingly long hours in factory sweatshops, leaving them with little time for walking around the city streets. The main working-class experience of the collapse of the gender order during World War I was thus formed not in the public spaces of Austrian cities, but in the closed workshops and halls of the Austrian wartime industry.

Gender Diversification of the Working Class and the Search for New Horizons

At the beginning of the chapter, I noted how immediately prior to and during World War I the social status of female reproductive work was transformed significantly, changing the prewar gender order in many ways as well. The reassessment of housework and cooking made women into publicly active persons and reshuffled family hierarchies. The other side of this process, then, was the strengthening of the emphasis on the differences between genders rather than on their equality. Women became public people, who in certain situations could even make decisions for their male counterparts based on the knowledge available to them, yet within traditional masculine aspects of everyday life, such as when and how to eat.[62] However, working-class women did not achieve these changes as men had—through productive work—but rather through a significant rise in the value of reproductive work, which became a prominent public topic in the context of wartime shortages.

On the eve of World War I, it was thus still possible to relegate women to the domestic sphere, and the constant emphasis on women as mothers and housewives supported the broader argument of women's weakness, fragility, and the need for their special protection, which significantly influenced the world of manual labor.[63] Because of the perception of their "natural weaknesses," women were generally seen as more worn down by industrial labor than their male counterparts. Within this context, an experiment was conducted as part of research on industrial hygiene in 1912 in Frankfurt am Main, which Jakob Tanner refers to in his work, and which became renowned all over Europe on the eve of World War I. Lab animals were given various industrial poisons and their bodily abilities to deal with them were closely observed. The results were supposed to reveal which factory operations, and for how long, the human organism could survive in. Lab animals were selected to represent men as well as women. Dogs were used instead of men, while lab cats represented working women. Dogs thus represented physically strong men, while cats represented the more fragile fairer sex. The research results concluded that the lab dogs, and therefore working men, are more resilient against most industrial poisons and may therefore be used in various production operations in greater quantities and for a longer time.[64]

This complex discourse on industrial labor cast doubts on women's competence as providers of anything other than reproductive work. Although many women were employed in the prewar, and even more so in the wartime, industry, their position was generally different from their male counterparts. The amount of energy that they could spend for the benefit of their employers was smaller than men's, which corresponded to the amount of food that women's bodies were supposed to consume. In the first chapter, I mentioned the research of the Munich professor Carl von Voit, who, based on his empirical measurements, established the necessary daily amount of 2,691 calories for an average working man, and only 2,153 calories for a working woman. Nutritional experts throughout Europe published similar results, and further research generally replicated the basic androcentric perspective on nutrition as well as industrial labor. In the 1880s, the German economist and statistician Ernst Engel, for instance, arrived at mathematical calculations of average amounts of consumption of individual foods per sex and age, and concluded that a women must eat 14 percent less than a man. He based his findings on the basic unit of man as a "whole person" (*Vollperson*). So, in his research, women and children appeared only as the derivative of a male unit.[65]

During the period of European industrialization in the second half of the nineteenth century, the working man became the norm for any consideration of industrial labor or diet. His prominent position in the world of industrial labor was gradually cemented through the scientific research of various disciplines. The results of nutritional and physiological research also placed various demands on both sexes in terms of providing industrial labor on the one hand, and in terms of both sexes' demands regarding food consumption and basic material conditions on the other hand. The greater amount of work that the man conducted authorized him to receive a greater caloric intake and more meat in his daily diet. Physically demanding work, and the nutritional standards it entailed, was deeply entwined with the working-class form of masculinity and was an integral part of the identity of the androcentric working-class collective.

Beginning in the fall of 1914, however, the androcentric identity of the working class was dealt even greater blows in the production halls of Austrian factories than it was forced to accept in the everyday world of the streets of Austro-Hungarian towns. In the previous chapter, we noted how the reorganization of the Austrian wartime industry significantly transformed the working-class position within the hierarchy of the factory. In his contractual relationship, a worker became the object of the discourse of rationalization and of the disciplinary practices that took away almost all of his autonomy. The new organization of the inner hierarchies of the enterprises and the relevant laws divested him of his competency to actively take part in the decision-making processes on the amounts and the manner of the work that he provided for his employers.

This broad reformulation of the relations in the vertical hierarchy within the enterprise was mirrored in the horizontal level, i.e., in the relationships between the workers themselves, and it was precisely these relationships that had a noticeable gender aspect. While in the Austrian industry the positions of rationalization experts, factory owners or their military administrators continued to be assumed only by men, the working-class collective itself was transformed by the massive influx of working women. This influx, however, was not gradual, but extremely swift. The dynamic with which many previously male professions and work situations were taken up by women did not allow much time for gradual adaptation. Only several months after the outbreak of war, in the fall of 1914, official government bodies stated that "women's involvement in wartime production not only takes place in the current, primarily women's professions, but it is increasingly necessary to involve women even in positions otherwise reserved for men."[66] At the same time, there was an increase in notifications that women more often work the night shifts that were also primarily reserved for men before the war.[67]

Women thus found themselves in professions that were previously given to men, as well as in industrial branches, in which women's presence would have been unthinkable even right before the war. Although, for example, the ratio of women employed in the heavy iron and machine industries during the war did not surpass one-fifth and was therefore still small from an overall perspective, its rapid increase compared to the prewar years cannot be overlooked. In 1914, for instance, the entire Pilsen district declared that only 2.4 percent of the workforce in the iron and machine industries was female,[68] while during the two following years, this amount grew to almost 20 percent.[69] Experts estimated that the overall number of women employees in the Austrian economy of the war years grew to roughly one million people.[70]

Women in Czech factory halls did not always have to play a decisive role from a quantitative perspective, but even here their numbers in the workplace were many times higher than before the war. The almost ten times higher number of women employed in traditionally male branches, such as heavy machinery, could not have gone unnoticed by male workers, and further deepened the crisis of their male identity. Women's entrance into professions that were formerly the domain of men further cemented the ruling discourse that relegated workers to positions of mere tools of production that could be replaced at any time.

Thus, for example, the Austrian trade inspection reported for 1916 that "in the district of Vienna V, more and more women are called to highly qualified work in iron and munitions enterprises, and even in chemical plants. Similar conditions can be seen in the districts of Prague II, Trutnov, Pilsen, Pardubice and Brno."[71] In the Pilsen Škoda factory, women were entrusted with servicing heavy cranes,[72] and in the district of Prague II, which included Vinohrady, Smíchov, Žižkov, Benešov and Vlašim, women workers worked with autogenic welding machines and serviced one-ton minting presses, sledgehammers, or lathes.[73] In the munitions plant in the Prague neighborhood of Žižkov, the absolute number of women workers was higher than the male workforce,[74] and in the very center of Prague, women serviced heavy steam mills in two alternating shifts with no men present.[75] The central army administration had an eminent interest in the increased number of women in traditionally male professions, which were gradually experiencing a lack of military recruits. The official instruction of the Litoměřice military headquarters, for example, encouraged the military administrators of Czech factories to "take utmost care of their [male workers'] replacements and urge military administrators to put a special emphasis on this issue."[76]

The traditional component of the definition of prewar masculinity, i.e., qualified, relatively well-paid work, lost its potential to serve as the basis of positive male self-identification during the years 1914–1918. However, it was not just the presence of women in the previously male environment of the production halls. The legal framework of the working-class social security, which helped to cement the male primacy in providing industrial work since the 1880s, also went through significant changes.[77] According to law no. 33 from 1888 on mandatory health insurance, the employees of Austrian factories were separated into groups of men and women, and the support that these groups received in the event of illness or accident was calculated separately for each.[78] Thus women, in complete accordance with the ruling energy paradigm of industrial labor, received, as a workforce that expends a smaller portion of energy, 30 to 50 percent less on average than men. This calculation of health insurance benefits ensured that almost no woman would get a greater sum than a man in the event of illness, even in the few cases when she received similar pay.

The amendment of this law from January 9, 1917, however, cancelled this explicit division of health insurance benefits. Instead, it introduced eleven income classes, which then determined the prescribed daily financial benefit that was provided for the worker in the event of illness.[79] The criterion of gender for the calculation of the health insurance benefits completely disappeared and, starting in 1917, there were occasions in the Habsburg Monarchy when a woman received the exact same, or even higher, health care benefit as the man.[80] Since the rates of illnesses and accidents in the Austrian wartime industry were, as a consequence of the often dire conditions in production, significant and the number of insured working women constantly rose, these situations actually occurred quite often.[81]

The high rates of illness and accidents in the Austrian wartime industry, however, had a broader influence on the gender identity of male workers. After a series of serious work injuries, the Pilsen Škoda factory was forced to implement new women's work attires that were not too different from the men's clothes. Female crane operators and founders wore pants instead of skirts and tighter shirts instead of looser blouses. Loosely tied head scarves were replaced with scarves that covered the whole head.[82]

According to the reports of Austrian doctors, these hard-working women were most affected not only by short-term illnesses or work injuries, but also by physiological disorders that cast doubt upon their very gender. For instance, in 1917 the official health statistics revealed that roughly one-fourth of all Austrian women of reproductive age had ovarian disorders, and thus lower reproductive capabilities, due to exhaustion and malnutrition.[83] When women's main, and in many cases only, social goal was giving birth to children and raising them, these women workers actually lost their female essence.

Women in pants and men's work shirts, with short and covered hair, who due to malnutrition lost not only their reproductive capabilities, such as childbirth and breastfeeding, but who also often suffered the loss of secondary gender signs, ceased being seen as merely a temporary substitute for drafted male colleagues, but were sometimes understood as an entirely new gender that significantly invalidated the current gender order.[84] The transformation of working women's bodies as a result of wartime suffering and their intensive participation in the sphere of industrial production invalidated the very definition of biological sex and the hierarchy based on it. Men and women's entirely new experiences of the extreme conditions of wartime shortages at the home front signified the overturn of existing gender-based defining categories.[85] Many women now shared with men the cultural and physical attributes that before the war were reserved only for men, and, in fact, it was not always possible to distinguish them from men.

The upheaval of the prewar gender order was further energized by the unprecedented loosening of sexual morals. There was a rise in the number of women actively seeking a sexual partner in the same way a man would, and many working women kept adulterous relations not only with their male colleagues, but also with prisoners of war, who were sometimes used for various jobs in Habsburg factories.[86] Thus, for instance, in the spring of 1916, the military administrators of all the factories in northern and western Bohemia were asked to initiate extensive inspections, which were supposed to confront the rising spread of venereal diseases, most importantly gonorrhea and syphilis, among the labor force. These inspections did not distinguish whether the potential source of infection was a male or a female worker. In the eyes of the military administration, both sexes were equal in this respect.[87]

This almost ubiquitous collapse of the gender order understandably elicited sharp reactions, which not only could not reverse it, but also often encouraged it even further. A striking example of such a counterproductive reaction was the introduction of a highly disciplinary regime in several workers' colonies. Women were separated from men and guarded by a nightwatchman, whose task was to stop any sexual contact between the sexes. Their contact with the outer world was strongly limited, as they had to request permission to go out and were often bullied by the nightwatchmen.[88]

Women workers were thus seen as active agents who sought out sexual partners by themselves, and who therefore had to be properly disciplined.

Women in working-class housing colonies were actually submitted to a very similar disciplinary regime as the one the army used with soldiers in barracks. There, too, the soldiers' time was organized between training and rest, and any contact with the outside world was strictly controlled. The rationale for this was the presumed sexual unrestraint of the draftees, and the health and social danger that stemmed from this. In other words, seen through this perspective, there was no difference between a soldier and a working woman.

In the eyes of the male workers, many working women did not just adopt their clothes, jobs, or public authority, but their male sexuality as well. Carpenter Vojtěch Berger summarized this in his diary entry for 16 January 1916: "[I]t is hard to find a real girl nowadays. Women are beasts and many are worse than men. The war is spoiling the women and girls … men are on the battlefield and they are going out with someone else and these bitches always have to have a man on top of them."[89]

In the sphere of industry, World War I overturned the deep-rooted system of male values, based on the provision of qualified work and the perception of a man as an active subject of public affairs and the initiator of sexual contacts.[90] This moment has not been overlooked by contemporary gender studies. As R. W. Connell writes: "A wide range of responses can be made to the undermining of the bodily sense of masculinity. The one thing none of the men can do is ignore it. … Emphasizing the masculinity of industrial labor has been both a means of survival in exploitative class relations and a means of asserting superiority over women."[91]

Even the workers at the home front of World War I could not ignore that the central point of their masculine identity was cast into doubt. The deep restructuring of the working class aroused various reactions, which actually helped organized labor stay alive. As previously stated, when many of the workers' clubs and political parties that claimed the right to articulate working-class interests were suspended, the labor unions became the main institutional base of working-class culture. It was there, especially on the turf of the still functioning large unions, that steps were formulated on how to deal with the changes in the field of gender relationships.

Labor union functionaries' reactions to the dynamic and deep transformation of the gender order in the sphere of work were multifaceted. Union offices that previously united typically male professions tried for a long time to keep their membership and themselves deeply rooted in the stereotype that equated certain professions with the masculine identity. Union newspapers thus often printed highly emotional texts that were meant to strengthen the prewar male identity structured around strenuous, qualified work, such as the *Kovodělník* in February 1917:

> We would like the reader to consider a coal mine. An engine room. A machine worker commands a machine with purposeful movements of various levers that he operates with both hands and feet, and the big iron giant obeys to them. His shining limbs cast light about the engine room, rhythmically, in accordance with the movements of the giant metal limbs. ... The machine worker, while manipulating it, gazes steadily at the "scale," a small picture of what is going on in the mine shaft. He observes on the "scale" which of the cages is nearing the end of its track; one sits below at the bottom, while another rises to "the day" at the same time. ... Everything that takes place in the engine room is the center of the whole work process, beginning with the moment when the miner sets his hoe against the coal seam and begins to "scratch"—until the moment when the fully loaded coal cart leaves the mine. ... May labor be honored![92]

Work was thus linked to typically masculine attributes, such as aesthetic physical strain or the complexity and responsibility of male work processes, in this case manifested in having control over a heavy machine and handling the latest technology in the form of the graphic "scale."

Next to this emphasis on the masculine aesthetization of work, many union members also tried to disrupt the ruling discourse of heroism, which was primarily reserved for soldiers. Thus, labor unions often assumed the role of a similarly indispensable segment of the Austrian war effort as the army, paying the ultimate sacrifice at the front. However, this resonated more within the working-class collective itself, rather than outwardly, and did not have much of an influence on the occupation of the public space by the figure of the hero-soldier. Accordingly, union leaders adopted the communication strategy of comparing industrial work with military duty at the front predominantly when speaking inside of the working-class collective, such as in various internal publications and annual reports. In these materials, workers were often placed on a similar and in several aspects even on a higher level as heroic frontline fighters: "The assurance of Messrs. Sieghard or Škoda is not necessary. The true heroes of the home front also know that everyone should tip their hat even to the last worker. For it is they who have aided the war through their work in the munitions plants; without their labor, it would not be possible to wage war."[93]

The other side of this strategy of sticking to the masculine, heroic character of qualified, industrial work was relegating women to unqualified professions, and thus outside of the organized working-class collective. According to many union leaders, women lacked an organizational spirit that prevented them from actively joining the working-class collective, which threatened the very existence of the working class. As the largest Czech metalworkers' labor union stated: "Women ... were forced to take the place of their men, who were drafted into the war, to face, without much consideration, work in the factories in order to buy bread for their families and themselves. Not having enough organizational sense and expecting to return to their families soon, they don't think of uniting, of an orderly and purposeful defense of their interests. Women's labor progressed to great proportions during the war, and thus endangered orderly life and the safety of future generations."[94] In the eyes of union leaders, women's involvement in the Austrian wartime industry significantly threatened the ability of working-class collective action. Women were at the same level as children, old people, and invalids. Their rapid employment in wartime production paralyzed organized workers' struggles, so that employers who profited unfairly from workers' labor "could remain calm."[95]

The perception of women as emotional, as opposed to men as rational, with natural, spontaneous reactions, as opposed to men's collective organized behavior, and the inability to control women's actions eliminated them from the working-class collective. Built on male workers' self-identification as rationally behaving individuals with the proper outlook, education and collective culture, this self-identification of many male workers culminated in the organized political movement of Social Democracy.[96] According to the opinion of its male leaders, every member of such a movement had "a sense of discipline, did not yield to instantaneous ideas and moods, but relied on his knowledge of political events and knew their importance, thought about resistance, but first considered its probable outcome."[97] Women's behavior, which in the eyes of the male functionaries oscillated between extremes of absolute apathy and incomprehensible, emotional outbursts of dissatisfaction and unorganized resistance thus represented the antithesis of the politically aware worker, and was often referred to as being outside of working-class culture.[98]

Beneath the official statements of the labor unions denying female industrial workers membership in the organized working class, a fundamental dispute over the very nature of the organized workers' collective was taking place in several labor unions. The official language of organized labor continued to emphasize the masculine character of heavy industrial work, but the everyday reality in many production plants, where men worked side by side with women, raised urgent questions about the possibilities of preserving the prewar, androcentric identity of the working class. The litmus test in this case was the question of the acceptance of women as full members of typically male union organizations, which became one of the main issues in almost all of the big union associations. The unions had experience with employing women in auxiliary positions as secretaries, typists or accountants, but the massive influx of working women as full members of union organizations with the right to vote during conventions would entail a radical paradigm shift for many members.

Despite the fact that a wave of women's involvement affected many branches beyond just heavy industry right after the outbreak of the war, and escalated further during the year 1915, many labor unions refused to accept the new female employees as full members for quite a long time. The leadership of the Union of Railway Workers, for instance, managed to block the proposal to accept new female railway workers as members for almost the entire duration of the war. With the help of various administrative procedures, dealing with new female conductors' membership request submissions was postponed until the fall of 1917. After three years of waiting, this key point was finally on the agenda and a long and heated debate took place, in which, according to the official record "all board members participated in a very heated way."[99] The conclusion of this debate, however, was no different from the previous conclusions reached when this question was discussed—the problem of accepting women as full union members was postponed with the explanation that such an important decision can only be accepted at the meeting of the broader leadership of the union.

This meeting took place three months later, in December 1917. Since it was the last instance in the decision-making process, and, after three years of delays, it was not possible to postpone it any longer, the extended union board was forced to make an official decision. This was formulated on December 15, 1917 and clearly stated that the union should "limit the acceptance of women as union members where possible and avoid their interests."[100] Upon being made public, this decision was immediately met with an unexpectedly strong outrage of female railway employees and even several union leaders, so that one month later, the union was forced to significantly alter it, and state that, as to the question of women, "due to the fact that the statutes allow it, a resolution has been passed to accept them."[101] It is significant that this final decision was made by the same smaller board that five months earlier still refused to comment on the whole matter and referred it to the extended board.

The swiftness with which the railway union leaders' position on such a significant matter as the full membership of women employees changed illustrates the influence of the war on the thinking of the working-class elite. The unexpected breadth of the war catastrophe disrupted the sense of direction of the prewar period, and forced many union members to make far-reaching decisions in entirely new and unknown conditions. These decisions then quickly changed the shape of the working class. The leadership of the railway workers' union refused to accept women as full members in December 1917, but in March 1918, a union meeting was held in Moravská Ostrava in which over one hundred women were present.

According to the report of the supervising state body, they even constituted the majority of those present.[102]

In the first years of the war, the metalworkers' union blocked not only the acceptance of working women into the unions of the individual production plants, but also constantly lobbied military administrators and factory owners to limit, or to cease entirely, the hiring of female employees. When the metalworkers' leaders were not able to push through this agenda, they ceased the effort in the spring of 1916, sadly stating that "many women and invalids are hired in metalworks due to the shortage of eligible workers. It was impossible to stop this measure."[103] After this lament over new female competitors in the workforce, just like with the railway workers' union, the new female employees' path to full membership in the metalworkers' union was unexpectedly swift. The union not only formally opened itself to women in 1916, but it also actively sought out women through targeted campaigns.[104] Resistance on principle transformed into absolute affirmation practically within a month.

Once the Union of Czech Metalworkers actively opened itself to women 1916, it managed to relatively quickly stem the decrease in membership that had started to threaten its very existence in the first years of the war. From the fall of 1914 to the end of 1915, it lost half of its members.[105] Beginning in 1917, however, thanks to its successful recruitments among new female employees, union leaders could once again be satisfied not only with the fact that the decline in membership had stopped, but also with the membership growth in certain local union branches, which regained their prewar numbers thanks to the women members. Raising the number of members also had a significant financial aspect. While the union was in a dire financial situation until the year 1916, by mid 1917 its reserves had more than 182,000 Austrian crowns, which was sufficient enough to cover all of its expenses.[106]

Newly employed women also slowly joined the core of the organized working-class collective. In the second half of the war, not only was it no longer rare for women to be full members in many labor unions, but they were also involved in the activities of the top organization of the working-class movement—organized Social Democracy. The representatives of women union members were often invited to party meetings, where they were given space to articulate their demands. When, for example, the social democratic meeting took place in Brno in April 1917, Františka Skaunicová spoke alongside traditional political leaders. Her speech contained points that would be impossible to speak aloud in a closed male collective, and that openly articulated the radical request of women's total emancipation:

> Our men must withstand much suffering on the battlefield and their women must work in their stead at the home front and raise their children alone. Therefore the woman deserves the same rights as the man. Just like the man protects his homeland, the woman protects the family. ... But she is not rewarded for it and she is ignored everywhere. It would look a lot different if women would make the decisions even in the affairs of the municipality, the province and the state. They could apply their economic experiences to the questions of supplies. If they could speak in parliament, then international squabbles could be solved without bloodshed. When people talk everywhere about what must be endured, women must submit themselves to these talks, too, and endure in the fight for their rights until the time when they will be equal to men.[107]

We can thus conclude that workers generally adapted to the massive influx of female colleagues during World War I, but the adaptation was in no way smooth. Resistance to the gender diversification of the working class was significant and lasted until the end of the war for many workers' leaders. On the other hand, the example of the metalworkers' union reveals how gender diversification could help preserve the working-class organizations. Accepting women thus helped preserve the institutional base of working-class culture.

The more obvious it became that the world war would not be just a temporary derailment of the status quo, the more the transformation of gender relationships was understood as permanent. Just like in the case of industrial labor, the last two years of the war witnessed considerations of postwar reconstruction pointing to a new horizon, rather than to the renewal of the old order.[108] In terms of gender relationships, this meant that even male workers gradually conceded the irreversibility of a large part of the changes brought about by the war: "The men will return from their guns and canons to their jobs. The world will return to its previous path. Will everything be the same as before? In many ways yes, but in many was not at all. The war, the terrible war, besides the terror it has brought also broke the old orders, and raised up many forces that had previously been underestimated. One of the most important consequences of this war is the elevated significance of women,"[109] stated the monthly *Zájmy žen* in 1917, prepared by women editors but published under the patronage of the Czech Labor Unions Federation.[110]

The transformation of the prewar gender order, which affected workers profoundly, thus deconstructed the previous androcentric identity of the working class.[111] It became more and more difficult in the years 1914–1918 to equate one's status as a worker with a sense of model masculinity, for similarly to the amount of food or the scientifically prescribed quotas of fatigue, the rations of manhood for workers at the Austrian home front were significantly curtailed. The loss of certain competencies in the running of the household, the collapse of symbols of male public authority, and the sudden and extraordinarily far-reaching invasion of women into the sphere of industrial labor irreversibly destabilized male self-identification during the war years. The androcentric substance thus gradually disappeared from the catalogue of the defining components of the working class, which increasingly transformed itself into a gender diversified collective.

The Czech working class during the second half of World War I certainly did not become a collective based on the equal rights of the two sexes. For the majority of the male workers, equal rights for women were more a necessity rather than a desire. Nevertheless, World War I meant a radical and rapid paradigm shift for most male workers. In a short period of time, the Czech working class lost its androcentric identity and became a project that counted on, or had to count on, a mixed gender composition.

Notes

1. Bad Religion, "Heaven is Falling," *Generator*, Epitaph Records (1992).
2. Adina Lieske, *Arbeiterkultur und bürgerliche Kultur in Pilsen und Leipzig* (Bonn, 2007), 370–85.
3. On the person of Max Mack, see Michael Wedel, ed., *Max Mack: Showman im Glashaus* (Berlin, 1996).
4. Helmut H. Diederichs, *Frühgeschichte deutscher Filmtheorie. Ihre Entstehung und Entwicklung bis zum Ersten Weltkrieg* (Frankfurt am Main, 1996), 58–61.
5. Marie Bahenská, Libuše Heczková and Dana Musilová, eds., *Ženy na stráž! České feministické myšlení 19. a 20. století* (Prague, 2010); Jana Malínská, "Volební právo žen do říšské rady, českého zemského sněmu a obcí v letech 1848–1914," *Střed/Centre* 1 (2009): 24–57; Jitka Gelnarová, "Matka Praha a dcery její: diskuse o ženském volebním právu do pražské obce v občanském a dělnickém ženském hnutí mezi lety 1906 a 1909," *Střed/Centre* 2 (2011): 34–58; Libuše Heczková and Kateřina Svatoňová, "Úvod: Nebezpečná Božena Viková-Kunětická," in *Jus Suffragii. Politické projevy Boženy Vikové-Kunětické z let 1890–1926*, eds. Libuše Heczková and Kateřina Svatoňová (Prague, 2012), 7–19.
6. *Nová Doba* (September 10, 1914): 3.
7. Ibid.
8. Ibid.
9. Marie Bahenská, "Pomalu, pozvolna, po špičkách. K chápání a reflexi pojmu emancipace v českých zemích v 19. a 20. století," *Moderní dějiny*, suppl. 1 (2008): 444–57.
10. Cf. Nancy Folbre, *Greed, Lust and Gender: A History of Economic Ideas* (Oxford and New York, 2009).

11. On the concept of sepharate spheres, see Karin Hausen, "Family and Role Division: The Polarisation of Sexual Stereotypes in the Nineteenth Century: An Aspect of the Dissociation of Work and Family Life," in *The German Family: Essays on the Social History of the Family in Nineteenth and Twentieth-Century Germany*, eds. Richard J. Evans and Robert W. Lee (London, 1981), 51–83; Linda Kerber, "Separate Spheres, Female Worlds, Woman's Place: The Rhetoric of Women's History," *Journal of American History* 75, no. 1 (1988): 9–39; Leonore Davidoff and Catherine Hall, *Family Fortunes: Men and Women of the English Middle Class 1780–1850* (London, 2002); Joan Landes, ed., *Feminism, the Public and the Private* (Oxford, 1998); Ute Frévert, ed., *Bürgerinnen und Bürger. Geschlechterverhältnisse im 19. Jahrhundert* (Göttingen, 1988). For the Czech context: Daniela Tinková, *Tělo, věda, stát: zrození porodnice v osvícenské Evropě* (Prague, 2010); Alena Šimůnková, "Statut, odpovědnost a láska: vztahy mezi mužem a ženou v české měšťanské společnosti v 19. století," *Český časopis historický* 95 (1997): 55–107. On the general gender history of World War I in Austria-Hungary, see most recently Christa Hämmerle, Oswald Überegger, and Brigitta Bader Zaar, eds., *Gender and the First World War* (Basingstoke, 2014); Christa Hämmerle, *Heimat/Front. Geschlechtergeschichte(n) des Ersten-Weltkriegs in Österreich-Ungarn* (Vienna, 2014).
12. Kirsten Schlegel-Matthies, *Im Haus und am Herd. Der Wandel des Hausfrauenbildes und der Hausarbeit 1880–1930* (Stuttgart, 1995), 20–78.
13. Gabriele Czarnowski and Elisabeth Meyer-Renschhausen, "Geschlechterdualismen in der Wohlfahrtspflege: Soziale Mütterlichkeit zwischen Professionalisierung und Medikalisierung," *L'Homme. Zeitschrift für feministische Geschichtswissenschaft* 2 (1994): 121–40.
14. Christoph Sachße, *Mütterlichkeit als Beruf. Sozialarbeit, Sozialreform und Frauenbewegung 1871–1929* (Weinheim, 2003), 94–120.
15. Kejřová, *Dělnická kuchařka se zřetelem na malé dělnické domácnosti*, 5.
16. Anuše Kejřová, *Úsporná kuchařka: zlatá kniha malé domácnosti* (Hradec Králové, 1905).
17. Anuše Kejřová, *Kniha vzorné domácnosti: Vyzkoušené rady, pokyny a předpisy pro hospodyňky, jimž vzorné a úsporné vedení domácnosti na srdci leží* (Prague, 1916), 5.
18. *Nová Doba* (February 2, 1915): 3.
19. On the working-class families in the Bohemian lands, see Jiřina Svobodová, "Rodina a rodinný život pražského dělnictva," in *Stará dělnická Praha. Život a kultura pražských dělníků 1848–1939*, eds. Antonín Robek, Mirjam Moravcová and Jarmila Šťastná (Prague, 1981), 107–36.
20. *Zájmy žen. Časopis pro zájmy žen výdělečně pracujících* 1 (January 1, 1917): 2.
21. *Zájmy žen. Časopis pro zájmy žen výdělečně pracujících* 6 (June 1, 1917): 46 a 7; (July 1, 1917): 54.
22. Martin Jemelka, ed., *Lidé z kolonií vyprávějí své dějiny* (Ostrava, 2009), 413–18.
23. Article "Starosti v domácnostech", *Kovodělník. Orgán svazu dělnictva zaměstnaného výrobou a zpracováním kovů a drahokovů v Rakousku* 42 (October 18, 1916): 171.
24. Robert Marschner, *Die Fürsorge der Frauen für die Heimkehrenden Krieger* (Prague, 1916), 7.

25. Maureen Healy, "Becoming Austrian: Women, the State, and Citizenship in World War I," *Central European History* 35, No. 1 (2002): 1–35; Elizabeth Domansky, "Militarization and Reproduction in World War I Germany," in *Society, Culture, and the State in Germany 1870–1930,* ed. Geoff Eley (Ann Arbor, 1996), 437–42.
26. The historiography on war and gender is huge. The most important works from recent years include: Joanna Bourke, *Dismembering the Male: Men's Bodies, Britain and the Great War* (Chicago, 1996); Jessica Meyer, *Man of War: Masculinity and the First World War in Britain* (Basingstoke, 2009); Sabine Kienitz, "Body Damage: War Disability and Constructions of Masculinity in Weimar Germany," in *Home/Front: The Military, War and Gender in 20th Century Germany, eds.* Karen Hagemann and Stefania Schüler Springorum (New York and Oxford, 2002), 181–204; Deborah Cohen, *The War Came Home: Disabled Veterans in Britain and Germany 1914–1939* (Berkeley, 2001); Michael Roper, *The Secret Battle: Emotional Survival in the Great War* (Manchester, 2010); Santanu Das, "Kiss Me, Hardy: Intimacy, Gender, and Gesture in First World War Trench Literature," *Modernism/Modernity* 9, no. 1 (2002): 51–74; Paul Lerner, *Hysterical Men: War Psychiatry and the Politics of Trauma in Germany 1890–1930* (Ithaca, 2003); Elaine Showalter, "Rivers and Sassoon: The Inscription of Male Gender Anxieties," in *Behind the Lines: Gender and the Two World Wars, eds.* Margaret R. Higonnet, Jane Jenson, Sonya Michel, and Margaret Collins-Weitz (New Haven, CT and London, 1987), 61–69; Mary M. Roberts, *Civilization Without Sexes: Reconstructing Gender in Postwar France, 1917–1927 (*Chicago, 1994); Jon Lawrence, "Forging a Peaceable Kingdom: War, Violence, and Fear of Brutalization in Post-First World War Britain," *Journal of Modern History* 75, no. 3 (2003): 557–89; A. Belzer-Scardino, *Women and the Great War: Femininity in Italy* (New York, 2010).
27. Šedivý, *Češi, české země a velká válka 1914–1918,* 267. On childern welfare during the war, see further T. Zahra, *Kidnapped Souls: National Indifference and the Battle for Childern in the Bohenian Lands 1900–1948* (Ithaca, 2008), 79–105.
28. See the classical work by Georg Mosse: George L. Mosse, *Fallen Soldiers: Reshaping the Memory of the World Wars* (New York and Oxford, 1990). On wartime and early postwar masculinity, see further, for example: Nancy M. Wingfield and Maria Bucur, "Introduction: Gender and War in Twentieth Century Eastern Europe," in *Gender and War in Twentieth Century Eastern Europe,* eds. Nancy M. Wingfield and Maria Bucur (Bloomington, 2006), 1–20; Maureen Healy, "Civilizing the Soldier in Postwar Austria," in *Gender and War in Twentieth Century Eastern Europe,* eds. Nancy M. Wingfield and Maria Bucur (Bloomington, 2006), 47–69.
29. The most relevant works in this respect include Garreth Stedman-Jones, *Outcast London: A Study in the Relationship Between Classes in Victorian Society* (Oxford, 1971); Edward. P. Thompson, *The Making of the English Working Class* (London, 1961); Eric J. Hobsbawm, *Labouring Men: Studies in the History of Labour* (London, 1965); Jürgen Kocka, *Arbeitsverhältnisse und Arbeiterexistenzen, Grundlagen der Klassenbildung im 19. Jahrhundert* (Bonn, 1990); Jürgen Kocka, *Klassengesellschaft im Krieg. Deutsche Sozialgeschichte 1914–1918* (Göttingen,

1973). For a good summary, see Thomas Welskopp, "Atbeitergeschichte im Jahr 2000. Bilanz und Perspektive," *Traverse* 2 (2000): 15–31; Bettina Hitzer and Thomas Welskopp, eds., *Die Bielefelder Sozialgeschichte. Klassische Texte zu einem geschichtswissenschaftlichen Programm* (Bielefeld, 2010).

30. In the Czech context, see, above all, Fasora, *Dělník a měšťan.*
31. Raymond C. Sun, "Hammer Blows: Work, the Workplace, and the Culture of Masculinity Among Catholic Workers in the Weimar Republic," *Central European History* 2 (2004): 245–71; Gregory L. Kaster, "Labor's True Man: Organized Workingmen and the Language of Manliness in the USA, 1827–1877," *Gender and History* 1 (2001): 24–64; Paul A. Custer, "Refiguring Jemima: Gender, Work and Politics in Lancashire 1770–1820," *Past & Present* 195 (2007): 126–58.
32. These exceptions are: Rudolf Kučera, "Marginalizing Josefina: Work, Gender and Protest in Bohemia 1820–1844," *Journal of Social History* 46, no. 2 (2012): 430–48; Jan Mareš and Vít Strobach, "Třída dělníků i žen? Proměny chápání genderových vztahů v českém dělnickém hnutí (1870–1914)," *Střed/Centre* 2 (2012), 34–68.
33. Healey, *Vienna and the Fall of the Habsburg Empire,* 264–66.
34. Ibid., 263–64.
35. On the concept of hegemonic masculinity, see Lothar Böhnisch, *Männliche Sozialisation: Eine Einführung* (Weinheim, 2004); Raewyn. W. Connell, "The Big Picture: Masculinities in Recent World History," *Theory and Society* 22, no. 5 (1993): 507–44; Raewyn W. Connell, *Masculinities* (Cambridge, 2005).
36. Healey, *Vienna and the Fall of the Habsburg Empire,* 266–70. On similar phenomena in the English context, see David Sylbey, "Bodies and Cultures Collide: Enlistment, the Medical Exam, and the British Working Class, 1914–1916," *Social History of Medicine* 17, no. 1 (2004): 61–76.
37. ACCP, Paměti Vojtěcha Bergera, Z Války, Kniha III., od 12. února 1917 do 26. ledna 1918, 49.
38. Joshua S. Goldstein, *War and Gender: How Gender Shapes the War System and Vice Versa* (Cambridge and New York, 2004), 252–301.
39. Ludmila Cuřínová, "Ústav pro národní eugeniku," in *Technokracie v českých zemích (1900–1950),* eds. Jan Janko and Emilie Těšínská (Prague, 1999), 151–56.
40. *Wiener Medizinische Wochenschrift* 15 (1916), 590–94, quoted in Healy, *Vienna and the Fall of the Habsburg Empire,* 267. On the person of Julius Tandler, see Karl Sablik, *Julius Tandler. Mediziner und Sozialreformer* (Vienna, 1983).
41. This no doubt sounds better in Czech than it does in English:

    Byl hezký hoch a jeho zraky se bystře na svět dívaly
    Na slunný kraj, na šedé mraky, jež honí bouře přívaly
    a mohutnou měl, silnou plec—Jan Broskovec
    Na šachtě dřel pro svoje pány, na šachtě kopal, nabyl sil
    že neznal svět a neznal hany, svým klidným tempem stále žil
    jak uzavřené ptáče v klec—Jan Broskovec
    A přišel čas, na vojnu vzali jej páni v límcích stříbrných
    mu modré gatě, bluzu dali a palaš jako padlý sníh
    A dragoun byl dobrý přec—Jan Broskovec
    Pak byla bitva, voje šedé se hnaly proti sobě vstříc

na těžkých koních četa jede, na blízkou smrt si nemyslíc
A stala bolestná se věc—pad Jan Broskovec

ACCP, Paměti Vojtěcha Bergera, Kniha II. od 15. ledna 1914 do 4. října 1919, 35–36.

42. *Kovodělník. Orgán svazu dělnictva zaměstnaného výrobou a zpracováním kovů a drahokovů v Rakousku* 34 (August 20, 1914): 1 (emphasis in original).
43. Habrman, *Mé vzpomínky z války*, 27.
44. Ibid., 33.
45. Robert Marschner, *Die Fürsorge der Frauen für die heimkehrenden Krieger* (Prague, 1916), 2.
46. ACCP, Paměti Vojtěcha Bergera, Z Války, Kniha I. 28. července 1914 do 24. června 1915, 18.
47. Ernst Kantorowicz, *The King's Two Bodies: A Study in Mediaeval Political Theology* (Princeton, NJ, 1957).
48. Peter Urbanitsch, "Mýtus pluralismu a realita nacionalismu. Dynastický mýtus habsburské monarchie—zbytečná snaha o vytvoření identity?," *Kuděj* 1 (2006), 46–68; and 2 (2006), 35–53.
49. Daniel L. Unowsky, *The Pomp and Politics of Patriotism: Imperial Celebrations in Habsburg Austria 1848–1916* (West Lafayette, IN, 2005).
50. See the pioneering work on the American case, Mark E. Kann, *The Gendering of American Politics: Founding Mothers, Founding Fathers, and Political Patriarchy* (London, 1999).
51. *Nová Doba* (July 31, 1914): 4.
52. *Zájmy žen. Časopis pro zájmy žen výdělečně pracujících* 2 (February 1, 1917): 13.
53. ACCP, Paměti Vojtěcha Bergera, Kniha II., od 15. ledna. 1914 do 4. října 1919, 53.
54. Sigrid Augeneder, *Arbeiterinnen im Ersten Weltkrieg. Leben- und Arbeitsbedingungen proletarischer Frauen in Österreich* (Vienna, 1987), 121–27.
55. *Bericht der k. k. Gewerbe-Inspektoren über ihre Amtstätigkeit im Jahre 1916* (Vienna, 1919), IX–XXII.
56. *Zájmy žen. Časopis pro zájmy žen výdělečně pracujících* 10 (October 1, 1917): 75.
57. Healey, *Vienna and the Fall of the Habsburg Empire*, 204–05.
58. *Právo Lidu* 216 (August 8, 1917): 3.
59. *Nová Doba* (Auguts 8, 1914): 3.
60. *Nová Doba* (January 8, 1915): 6.
61. Nedvěd, *Jak to bylo na českém západě 1914–1918*, 50.
62. Reinhard J. Sieder, "Behind the Lines: Working-Class Family Life in Wartime Vienna," in *The Upheaval of War: Family, Work and Welfare in Europe 1914–1918*, eds. Richard Wall and Jay Winter (Cambridge and New York, 1988), 114–15.
63. On the relationship between science and women's bodies, see Edward Shorter, *Women's Bodies: A Social History of Women's Encounter with Health, Ill-Health, and Medicine* (New Brunswick, 1991).
64. Tanner, *Fabrikmahlzeit*, 76–77.
65. Ibid., 77.
66. *Bericht der k.k. Gewerbe-Inspektoren über ihre Amtstätigkeit im Jahre 1914* (Vienna, 1915), CXXXI–CXXXVI.

67. Ibid., CXXXVI.
68. Augeneder, *Arbeiterinnen im Ersten Weltkrieg*, 55.
69. In 1916, Austrian authorities registered 19.4 percent of women among the workers of the Pilsen Skoda factory.
70. Wilhelm Winkler, *Die Einkommensverschiebungen in Österreich während des Weltkrieges* (Vienna and New Heaven, 1930), 32.
71. *Bericht der k.k. Gewerbe-Inspektoren über ihre Amtstätigkeit im Jahre 1916*, CV.
72. Ibid., CVI.
73. Ibid., CXIV.
74. Ibid., 221.
75. Ibid., 208.
76. CMAP, Fond 9. Sborové velitelství (1883–1919), karton 131, sign. 75-4/45.
77. On the respective laws see Miloslav Martínek, "Přehled vývoje rakouského zákonodárství v oblasti chudinství, zdravotnictví a sociální správy," *Sborník k dějinám 19. a 20. století* 4 (1977): 63–85.
78. RGBl 1888, 57–71. Here, above all, § 7, page 59.
79. RGBl 1917, 9–18. Here, § 7, page 12.
80. CMAP, Fond 9. Sborové velitelství (1883–1919), karton 115, č. res. 23101.
81. Vladimír Průcha, "Nástin vývoje nominální mzdy zaměstnaného průmyslového dělníka v Československu v letech 1913–1937," *Sborník historický* 13 (1965): 82.
82. *Bericht der k.k. Gewerbe-Inspektoren über ihre Amtstätigkeit im Jahre 1916*, CV.
83. Augeneder, *Arbeiterinnen im Ersten Weltkrieg*, 87.
84. On similar developments in Germany and Great Britain, see Elizabeth Domansky, "Militarization and Reproduction in World War I Germany," in *Society, Culture, and the State in Germany, 1870–1930*, ed. Geoff Eley (Ann Arbor, 1996), 427–64; Regenia Gagnier, *Subjectivities: A History of Self-Representation in Britain, 1832–1920* (New York and Oxford, 1991), 59–62.
85. Cathleen Canning, *Gender History in Practice. Historical Perspectives on Bodies, Class, and Citizenship* (Ithaca and London, 2006), 41–46.
86. Healey, *Vienna and the Fall of the Habsburg Empire*, 278–79. On the collapse of the current gender order during World War I, see further Dagmar Herzog, *Sexuality in Europe: A Twentieth-Century History* (Cambridge and New York, 2011), 45–61. Lisa M. Todd, "'The Soldier's Wife Who Ran Away with the Russian': Sexual Infidelities in World War I Germany," *Central European History* 44 (2011): 257–78.
87. CMAP, Fond 9. Sborové velitelství (1883–1919), karton 96, sign. 75-4/45-2. Instrukce vojenského velitelství v Litoměřicích správcům militarizovaných podniků z 18. 3. 1916.
88. Augeneder, *Arbeiterinnen im Ersten Weltkrieg*, 61–69. On wartime prostitution and the spread of venereal diseases, see further Nancy M. Wingfield, "The Enemy Withinn: Regulating Prostitution and Controlling Venereal Disease in Cisleithanian Austria During the Great War," *Central European History* 46 (2013): 568–98.
89. ACCP, Paměti Vojtěcha Bergera, Z Války, Kniha II., od 25. června 1915 do 11. února 1917, 108.

90. See the seminal study,Alice Kessler-Harris, "Treating the Male as Other: Re-defining the Parameters of Labor History," *Labor History* 34 (1993): 190–204.
91. Connell, *Masculinities*, 55.
92. *Kovodělník. Orgán svazu dělnictva zaměstnaného výrobou a zpracováním kovů a drahokovů v Rakousku* 7 (February 15, 1917): 25.
93. *Österreichischer Metallarbeiterverband. Bericht über die Tätigkeit des Verbandes in den Verwaltungsjahren 1914–1920* (Vienna, 1921), 11.
94. *Kovodělník. Orgán svazu dělnictva zaměstnaného výrobou a zpracováním kovů a drahokovů v Rakousku* 17 (April 26, 1917): 66.
95. *Österreichischer Metallarbeiterverband*, 18.
96. On the prewar Social Democracy, see Jiří Kořalka, *Češi v habsburské Říši a v Evropě 1815–1914. Sociálněhistorické souvislosti vytváření novodobého národa a národnostní otázky v českých zemích* (Prague, 1996), 228–75; Hugo Pepper, "Die frühe österreichische Sozialdemokratie und die Anfänge der Arbeiterkultur," in *Sozialdemokratie und Habsburgerstaat*, ed. Wolfgang Maderthaner (Vienna, 1988), 79–100; Harald Troch, *Rebellensonntag. Der 1. Mai zwischen Politik, Arbeiterkultur und Volksfest in Österreich (1890–1918)* (Vienna, 1991); Helmut Konrad, "Arbeiterbewegung und bürgerliche Öffentlichkeit. Kultur und nationale Frage in der Habsburgermonarchie," *Geschichte und Gesellschaft* 20, no. 4 (1994): 506–18.
97. Nedvěd, *Jak to bylo na českém západě 1914–1918*, 20.
98. See the claims that were raised by social-democratic women of Austria at their convention in October 1917. "Prezidium českého místodržitelství podřízeným politickým úřadům 16. listopadu 1917. Relace o sjezdu žen sociálně demokratické strany Rakouska pořádaném ve Vídni ve dnech 20–24. října," in *Sborník dokumentů k vnitřnímu vývoji českých zemí za 1. světové války*, vol. IV, document no. 85, 203–05.
99. LUAP, Fond staré odborové spolky, sign. 3011/229. Zápis z jednání schůze užšího představenstva Svazu železničních zřízenců ze dne 7. září 1917.
100. LUAP, Fond staré odborové spolky, sign. 3011/229. Zápis z jednání schůze širšího představenstva Svazu železničních zřízenců ze dne 15. prosince 1917.
101. LUAP, Fond staré odborové spolky, sign. 3011/229. Zápis z jednání schůze užšího představenstva Svazu železničních zřízenců ze dne 19. ledna 1918.
102. Zpráva moravského místodržitelství ministerstvu vnitra o sociálnědemokratickém shromáždění železničářů v Moravské Ostravě z 27. března 1918, Rudolf Neck, ed., *Arbeiterschaft und Staat im Ersten Weltkrieg*, document no. 470, 432–33.
103. LUAP, Fond staré odborové spolky, sign. 4385/317. Zápis z jednání schůze širšího představenstva Svazu českých kovodělníků ze dne 26. března 1916.
104. Ibid.
105. LUAP, Fond staré odborové spolky, sign. 4385/317. Zápis z jednání schůze širšího představenstva Svazu českých kovodělníků ze dne 15 srpna 1915.
106. LUAP, Fond staré odborové spolky, sign. 4385/318. Zápis z jednání schůze širšího představenstva Svazu českých kovodělníků ze dne 1. července 1917.
107. Zpráva moravského místodržitelství ministerstvu vnitra z 30. dubna 1917 o socialistickém mírovém shromáždění konaném 27. dubna v Brně, in Ru-

dolf Neck, ed., *Arbeiterschaft und Staat im Ersten Weltkrieg*, document no. 184, 283–84.

108. On this shift in the perception of the collapse of the old order, see Michal Pullmann, "Revoluce a utváření nového: Vídeň, Prague a Berlín kolem r. 1918. K hodnotovým aspektům tří revolucí ve střední Evropě," unpublished dissertation.
109. *Zájmy žen. Časopis pro zájmy žen výdělečně pracujících* 1 (January 1, 1917): 1.
110. LUAP, Fond staré odborové spolky, sign. 4385/317. Zápis ze schůze širšího představenstva Svazu českých kovodělníků z 26. března 1916.
111. Mareš and Strobach, "Třída dělníků i žen?," 34–68.

Chapter 4

Rationed Anger: The Politics of Protest

"He who complains will be needlessly kept alive by his discontent."
—Johannes Urzidil, "Karl Weissenstein"[1]

The Pilsen Slipper

Monday, August 13, 1917 was not a safe day for the respectable citizens of Pilsen to be out on the streets of their city. Pilsen had been hit by an escalating wave of unrest, and the town and its surroundings were under martial law. On August 13 and 14, the streets of Pilsen were awash in violent chaos, which deeply shook one of the main industrial cities of the Habsburg Monarchy. The wave of theretofore unprecedented violence was ignited by a seemingly insignificant incident that occurred around two o'clock in the afternoon during the sale of food rations. A line had been forming in front of a store since the early morning hours, and shortly after noon the number of waiting people had reached roughly 2,000.[2] Though the crowd was composed mainly of women, there were also several older children, adolescents, and adult men.[3]

This was not an unusual situation within the context of the wartime ration system. Long food lines of mostly, but not exclusively, women were a relatively common occurrence in many Austrian towns, and the state apparatus usually not only managed to contain any violent outbursts within tolerable limits, but also in many cases averted them. Food lines, moreover, had become the site for Austrian authorities to monitor popular moods. The presence of uniformed, but in many cases even secret, policemen helped to suppress angry outbursts and to track the situation, contributing to prevent street violence.

Since 1915, the local supply offices had been working closely with the police in several Austrian towns. The supply offices reported, for example, the amount of bread for sale and the anticipated number of buyers to the police. The police departments then subtracted the number of bread loaves from the number of people waiting and ascertained the necessity and level of police deployment in individual municipal districts.[4] Although some incidents could not have been prevented even in the first years of the war, the monitoring of popular moods and subsequent prevention of unrest in Austria in the initial war years managed to maintain a rather high level of order in the Austrian hinterland. When, for example, the Austrian government summarized in August 1916 the problems that had thus far taken place in the wartime industry, it concluded that only seven serious incidents restricting production for longer than a few hours had occurred, and it rated the overall situation in the wartime Czech industry as largely calm.[5] The number and strength of street protests and factory strikes declined during 1915–1916 to a minimum and the rationalized economic system, together with the mechanisms for monitoring public opinion and preventive police measures, basically guaranteed Austrian wartime production peaceful conditions.[6]

The year 1917, however, was a radical turning point when various local protests spun out beyond the control of the Austrian authorities throughout the whole state.[7] This is also what happened in Pilsen on August 13 and 14, 1917. On these days, too, the crowd was carefully monitored by soldiers who, if only by their presence, were there to deter violent excesses during the distribution of food. But when one of the soldiers took action against a woman who had thrown a bag of corn flour on the ground, he provoked the largest wave of protests and violence that Pilsen had ever known.[8]

Just as the soldier tried to apprehend the woman, the anger of the crowd spontaneously turned against him and threatened to become a physical confrontation. The soldier decided to retreat, but the quickly forming crowd could not be stopped. At first, a large number of women tried to loot a passing flour truck. The patrolling soldiers managed to avert this, but the situation spiraled completely beyond the control of the soldiers and the assisting police.

After the unsuccessful looting attempt, the growing crowd took off to Café Waldek (still one of the luxury hotels in the city today), where it threw small bags of corn flour at the guests so forcefully, that all of the people in the cafe ran away. Apparently, one of the guests at the cafe triggered this outburst when he stared mockingly at the protesters through the window. Then the crowd, already numbering between 3,000 and 4,000 people, went to the private apartment of the liberal mayor of Pilsen, where they broke several windowpanes. The police forced the crowd away from the mayor's apartment, but the flood of violence and looting did not abate. The large crowd divided itself into many smaller groups that crippled the center of town that afternoon. Passing cars, as well as many local stores, were looted. When the police or the army tried to intervene, the crowd would disperse only to reform itself again in another place and continue looting. Mere minutes after one group would scatter, more people would amass at a different place, so that the overall scope of the riot grew every hour and over the course of the afternoon the number of people participating in the rioting was estimated to be 5,000.[9]

The fast-moving crowd punished anyone it disliked. Several apartments, whose owners threw objects at the crowd from their windows, fell prey to rioting or coarse heckling. Roughly fifty grocery, clothing and women's hat and shoe stores were looted. According to an eyewitness: "And then … crowds of people all over Pilsen broke into shops and looted. People broke into unpopular stores as well as stores that nobody objected to. … They carried out any goods they could get their hands on randomly, nothing necessary or even useful to them. … People, who would earlier have looked for a policeman for over an hour to give him an old rusty key they had found on the street pushed their way into the store … and when they got inside they grabbed, for example, one slipper and then pushed their way back out to the street to victoriously wave the odd slipper over their heads."[10]

Towards evening some younger workers from the Škoda factory plant joined the crowd. The situation calmed down somewhat around ten o'clock at night thanks only to the deployment of a larger number of armed soldiers and sharp shooters. However, even this harsh intervention did not solve the problem altogether. Early the next day when Škoda workers returned from their night shift, more unrest occurred, this time at the Pilsen train station. Here, workers forcefully stopped trains to Prague and Vienna and threatened the passengers and crew.[11] When more and more train station employees joined in the unrest and the main western Bohemian railway junction was at risk of collapsing, it was no longer merely a local rebellion. Cutting off rail transportation in the direction of Prague or Vienna could potentially cause extensive economic damages and have strategic consequences for the overall military supply system. Therefore, Austrian authorities reached for their final option to quiet the town and officially declared martial law just before noon. Notices were plastered around the whole town and twenty squadrons of soldiers and two squadrons of hussars finally managed to restore order on the streets of Pilsen on the evening of August 14.[12] Although the subsequent investigation and arrests could not affect everyone who participated in the unrest, the local police were successful in arresting several dozen individuals, who were subsequently prosecuted. To a certain extent, the state powers understood the trials as an exemplary punishment of heretofore unthinkable riots. The sentences reflected this: the defendants were sentenced to months and even years in jail. Even the unnamed popular hero, who greeted his surroundings with the stolen slipper, did not escape. A Pilsen court sentenced him to four months in prison.[13]

The Uncatchable Subject of the Street Riots

The August events on the streets of Pilsen were not unusual in the context of the climactic year of 1917. Spontaneous protests, which were originally about shortages in food distribution, had become a common backdrop in Czech towns in the last two years of the war. According to the figures that Peter Heumos collected during his analysis of the wartime protests, the Austrian police recorded 252 cases of unrest in the Bohemian lands in 1917, which the local offices labeled as "hunger demonstrations." In several cases, however, these were rather massive unrests counting over 10,000 people, which often culminated in the declaration and strict enforcement of martial law.[14] Little changed the following year, when there were 235 reports of such riots by October 1918.[15] Together with a growing number of strikes, the waves of unrest, labeled as "hunger riots," were the most visible expression of the crumbling Austrian social consensus, which in the summer of 1917 led the Austrian government to consider whether it would not be better to stop the Monarchy's war effort on the Russian front despite its actual victory.[16] The "doses of anger", which drove thousands of people into the streets of many Austrian towns, dramatically increased especially in the last two years of the war.

The "hunger storms"—the most noticeable expression of this escalating anger—subsequently became a frequent object of historical research. Communist historiography often saw them as expression of popular dissatisfaction with the exploitative capitalist system and proof of the class struggle of the burgeoning urban proletariat.[17] The perspective that emphasized material impulses as the origin of mass protests in the last years of the war was still prevalent in Czech historiography even after 1989, but it paid more attention to national aspects.[18] The nationalist motives of the individual protests were pushed to the forefront and these protests were recast as an expression of anti-German sentiments, or as proof of the Czech-speaking population's gradual alienation from the Austrian state, which in its eyes was increasingly losing the legitimacy of its own existence.[19]

The common characteristic of the majority of the scholarship produced so far is a focus on the structural causes of the individual protests, which were then used to explain their development. The "hunger demonstrations," according to this perspective, originated as a reaction to the catastrophic food shortage, which was reflected in their form, often resulting in seemingly illogical explosions of spontaneous violent acts. According to a large part of the existing research, it was a form of protest that was qualitatively different from organized workers' strikes, which were also conducted by different participants. While the "hunger storms" were dominated, according to the recent research, by women or adolescent children, workers' strikes were carried out by an organized collective of male workers and were characterized by the rational negotiations of the strikers, who usually clearly formulated and presented their demands, negotiated with employers' representatives or the state and only rarely used violence. As Ivan Šedivý summarizes in his influential book on the Bohemian lands and World War I: "It is interesting that the hunger storms only rarely overlapped with the organized performance of the workers in the factories. The special composition of the demonstrators foreshadowed the outer appearance of these events: spontaneity, hysteria, crowd psychosis, tendencies toward unrestrained destruction and open violence."[20] In contrast, organized strikes and demonstrations, according to him, "initially erupted almost always solely out of economic and social reasons. Political demands, however, gradually began to appear."[21]

The relevant historiography thus, to a certain extent, assumes the perspective of the Austrian repressive bodies. They almost always deduced the character and form of the protest from the social and gender composition of the protesters. The division of protesters into women and adolescents on one side and men on the other at the same time was not random and reflected the paradigm of male rationality in opposition to female emotionality. Women were thus cast as being driven by biological tendencies toward irrational behavior, excessively succumbing to their natural urges and unexplained emotional explosions, often culminating in uncontrolled expressions of hysteria. Their biological essence was seen as less in line with adult men and more on the level of adolescent children. Women and adolescent children thus blended into one homogeneous group, whose behavior was contrasted to the behavior of adult men, characterized by rational thinking, clear formulation of their own interests and their subsequent collective action in order to fulfill them.[22]

This paradigm served as the basis for the differences in repressive practices, and Austrian security forces intervened differently against protesting women than against protesting men. When, for example, the Austrian police arrested the participants of the unrests that flared up around the Karlsgrube mine in the northern Bohemian town of Teplice in March 1917, the arrested persons were divided into men on one side and adolescents and women on the other. While the arrested men were handed over to the Austrian justice system, children and women were "enlightened and ordered to work."[23] The Austrian authorities thus fully respected the premise, according to which women and children were not full political citizens and as such could not bear full responsibility for any political behavior.

This division of the protesting crowd into rational men on the one side and emotional women and adolescent children on the other then influenced the interpretation of the protest events. The street unrests, often culminating in riots, were understood by the Austrian police and by most of the recent historiography as a typical expression of youthful or womanly imprudence and affected emotionality. The lack of understanding of the specific protest events, as for example the seemingly senseless looting of completely unnecessary slippers in Pilsen, was attributed precisely to the protesting crowd's missing rationality or directly to "crowd psychosis," which was understood as a characteristic of larger female collectives.

However, when we look more closely at the wave of street protests that affected the Bohemian lands in the last two years of the war, several nuances deviating from such a blanket assessment become apparent. In many cases, the protesters expressed more than mere dissatisfaction with the supply situation. Only a few were, by contrast, composed solely of a homogenous crowd of women or adolescents. For that matter, many men, who were present from the start, actively participated in the Pilsen riots described above, while others freely joined the protesting crowd as soon as their work shifts ended.

According to the police reports, on the afternoon of August 13, 1917, a rioting crowd began to loot stores in the center of Pilsen, where it had to overcome very solid barriers. All of the stores were not only securely locked, but many shop owners had installed heavy-duty bars on their doors and windows. The looting groups, however, overcame these metal bars rather quickly, which is difficult to do without male physical strength.[24] Throughout the night of August 13 and into August 14, the streets of Pilsen were at the mercy of violent groups composed of women and adolescents as well as workers returning from their day shifts. The next day, the workers even took over the protests when they occupied the Pilsen train station and violently stopped the departing trains.

The government sometimes saw Pilsen as one of the largest epicenters of wild unrest in the entire Habsburg Monarchy, but similar protests took place in many other large towns.[25] The then 18-year-old coal miner Josef Gurný recalled a similar wave of unrest at the beginning of July 1917 in one of the most important Austrian mining regions of Vítkovice, during which there were several fatalities:[26] "At the František shaft there was a miner, Stoklasa was his name, he led the looters in Přivoz. He had a metal baton in his hand, he broke the door. … There was a step there, he pried it out, the shutter flew open, he broke down the door and then straight into the shop. What was in there? Rare household things and clothes. … People were crowding in and taking whatever they could get their hands on."[27] The main participants of the wild unrests in Vítkovice were also a mixed crowd, which could scarcely have succeeded without the strength of the adult male workers. But even here the crazed looting did not have a strictly material purpose and, according to eyewitnesses, the looting crowd took practically everything that it could find with no regard to the usefulness or value of the stolen goods.

The waves of looting were also not always merely the acts of a wild crowd that spontaneously formed as a consequence of the material supply crisis. They were often the result of an organized and disciplined strike. Thus, for example, the steel and metal workers' strike in Kladno and Slaný at the beginning of May 1918 very quickly turned into raids by male bands against local shop and landowners, in which many were wounded and one person was killed.[28] The boundary between an organized, male workers' strike and unbridled street unrests was thus significantly blurred in many places and even the Austrian administration had to sometimes admit that the "hunger storms" were not just a phenomenon solely connected with the female temperament, but often contained a very significant male component.[29]

The seemingly incomprehensible and elemental unrests that Austrian authorities attributed to the predominance of women and children in the relevant protest groups were actually far more complex than mere explosions of hungry anger. The Austrian police and the middle-class observers shook their heads over the seemingly senseless incidents when the looting bands broke into the stores of otherwise popular shop owners in order to victoriously take a single slipper. Such moments did not match the common perception of "hunger storms" as a spontaneous wave of protests, whose only goal was the violent procurement of food. One slipper stolen in Pilsen could scarcely have improved one's material situation, nevertheless for its new owner it was a valuable trophy, which he showed off to everyone with pride.

Similar, seemingly grotesque forms of protest and looting pointed to vastly deeper patterns of dissatisfaction and revolt. Many of the actions of the protesting subject defy explanation in the economic categories that dominated the thinking of the Austrian police and other eyewitnesses, whose writings are the main bases of contemporary interpretations of wartime workers' unrests. The protesting crowd, composed of a mix of lower-class women, men and adolescents, was not just a passive object and it did not behave in accordance with middle-class understanding of the world, which was structured around the categories of economic logic.[30] When the vast majority of the relevant sources come from its opponents and suppressors, revealing the motives and practices of the protesting subject is thus quite difficult and must be approached very carefully.[31]

As a starting point, the classical concept of "moral economy," formulated by E. P. Thompson in the 1960s, seems worth exploring.[32] Looking at many Bohemian protests in greater detail, we can see how a specific non-market idea of justice and morality, so vividly illustrated by Thompson in the case of the English hunger riots in the eighteenth and nineteenth centuries, shaped the actions of the Bohemian protesting crowds during World War I. On the night of June 19, 1918, for instance, workers from the Laurin and Klement car company in Mladá Boleslav (Jungbunzlau) violently stopped a truck carrying eleven sacks of flour from Liberec (Reichenberg) after having waited all day for food. The workers put all eleven sacks of flour into the company store, which one of them had access to. Early in the morning of June 20, the sacks were moved to the factory kitchen, where the flour was distributed among other colleagues. After the flour was distributed, the workers who were responsible for the late-night hold-up of the truck arrived at the office of the district governor, where they admitted their deed. Their explanation of the nighttime act was quite simple: in exchange for their work, they were entitled to the flour anyhow and, by unloading it from the truck, they merely prevented the flour from being stolen by those who did not deserve it.[33]

The looting was a direct reaction to the catastrophic supply situation in Mladá Boleslav, where the local factories had received hardly any flour rations since the spring of 1918. The system of allocating ration cards had also completely collapsed and the workers, despite working at full capacity, could not buy any food, a fact that was even corroborated by the military administrator of the Laurin and Klement Company. The commandeering of the truck full of flour represented, in the eyes of the starving workers, basically a payment in kind, because the monetary payment they received for their work was practically worthless.[34]

In March 1917, the military headquarters in Prague reported a similar incident to the imperial court office in Baden near Vienna. According to this report, 600 people barged into a bakery in the Libeň neighborhood of Prague, where they stole 109 loaves of bread ready to be sold. However, the looting crowd did not make off with the loaves for free. In fact, it left money to cover a "fair price"—a price corresponding to the seller's moral right to make a reasonable profit for providing basic necessities in the eyes of the looting workers. However, this price was diametrically different from the price asked by the baker.[35] In this case, the workers applied their own notion about the fair price to be paid, one which the workers themselves did not object to. But because this fair price deviated greatly from the usurious price that the seller demanded, it had to be exacted by force.

We can generally conclude that the social practice of constantly waiting in line was a wholly unprecedented experience and a public humiliation that, coupled with a discourse of unjust suffering, in many cases generated spontaneous protest groups. These were often a reaction to the various supply troubles, but they did not disappear after the food was procured. In the last years of the war, the street as a social space and the food line as a social experience became the primary location of mutual communication between people who would otherwise not have come into contact with one another and for the cultivation of an alternative understanding of justice that went beyond the principles of market distribution. The subsequent protests then did not just put this "moral economy" into practice, but also often attacked the symbolic order of the liberal urban society itself. It was no coincidence that, just as in Pilsen in the summer of 1917, many popular protests marched into the central neighborhoods full of the "best addresses," where middle-class government workers, lawyers, and other representatives of the existing Austrian social order lived.

The bands rioting in Pilsen persistently regrouped in another place to continue looting after being dispersed. It was like a game of cat and mouse, where the looters used their perfect knowledge of the town to taunt the police and soldiers, who were composed to a great extent of hurriedly deployed Hungarian regiments and non-local policemen, who had little chance of capturing anyone while chasing them down the streets of an unfamiliar town.[36] In the summer of 1917, Pilsen experienced a two-day-long total upsetting of its spatial organization, when the otherwise peaceful middle-class streets of the city center were under the control of the lower-class working poor from the industrial suburbs, who duly demonstrated their dominance. Gaining control over the heart of the town symbolized a temporary, but no less definite, humbling of the main symbol of the entire social order. In the pre-industrial era, towns were one of the centers of power in a complex network of dominance, made up of other types of centers, such as monasteries or large feudal estates. However, during European industrialization, the town became the sole embodiment of a new, liberal order. Large towns represented concentrated power and an attack on their centers was also a symbolic attack on their political power as such.[37] The occupation of town centers as symbols of the reigning social order therefore meant a successful attack on the very essence of society.

When the efforts to overtake the city center were successful, the protesting crowd would erupt in spontaneous celebration, turning looting into an act of collective enjoyment. Enchantment stemming from an immediate feeling of immense power thus often devolved into the looting of goods that were, as many observers noted, seemingly "without necessity and usually without usefulness."[38] Heedless destruction, or the appropriation of entirely useless things, represented a direct negation of the middle-class liberal order in which there was no higher value than the economical, rational use of resources and the absolute protection of private property. The throwing away of looted food onto the streets for no reason, the squandering of rare corn flour that was used as ammunition against shouting onlookers despite the food shortage, or the victorious waving of a useless slipper above one's head were displays of just such a negation of the reigning liberal values and the expression of the specific rationality of the protesting crowd that did not operate within the categories of economic logic.

Moments that were more reminiscent of disorganized popular festivity than a dangerous wave of violence reveal that for many protesters, such riots were not just an opportunity to acquire badly needed food, but were also a departure from everyday, rationalized reality. When the majority of workers spent twelve or more hours at work in factory halls where they were required to make repetitious, rationalized movements, or when the majority of the protesting women stood in various lines almost all day on most workdays, the sudden disruption of middle-class hegemony over the urban spaces opened up new horizons in terms of understanding the urban space, as well as of the form of the specific events.[39] This is why the younger workers in particular brought elements of entertainment into the popular riots, for example by playing hide-and-seek with the police and transforming disorderly conduct and looting into acts of collective fun.[40]

If we remain in Pilsen, which in the last two years of the war practically embodied industrial unrest in Austria, and look at the urban development of the town before the outbreak of the war and during the war years in particular, we can see how working-class Pilsen became a laboratory of wartime economizing logic, leading to the total social and functional segregation of the urban space.

In Pilsen, the spatial division between the social classes first began, as in other Czech and Moravian towns, after the destruction of the town's fortifications, which in the western Bohemian metropolis took place in the first half of the nineteenth century.[41] A newly established ring of communal gardens grew up in the place of the original town ramparts and provided sufficient space for the representation of a new liberal order in the form of the opulent buildings of the municipal community center or the Pilsen theater, which were built on the outer boundaries of the town's center at the turn of the twentieth century. In parallel with the building of the architectural symbols of Pilsen middle-class culture, working-class housing developments began to be built primarily on the southern end of town in the second half of the nineteenth century.[42] The massive increase in the population of Pilsen especially after 1900 was accompanied by the rising spatial separation of the social classes. While the center of town had fewer apartments that adhered to the standards of middle-class housing, in the decade before 1914, the suburbs of Pilsen, such as Doudlevce, Lobzy, or Skvrňany, experienced an unexpected explosion of cheap rental houses, which provided shelter for the newly arriving working-class families.[43]

During the war years, these working-class housing developments were augmented with the building of lodging houses, where thousands of newly arrived Škoda workers were housed. The lodgings were simple, wooden row houses, mostly located in the Bory neighborhood near the Škoda factory. The enterprise's management limited the houses to the very minimum of necessities, so the mass working-class lodgings did not have any utilities and thousands of workers, who could not find any other place to stay in Pilsen, spent their nights in mass flophouses. Already at the end of 1916 and the beginning of 1917, there was a rise in complaints about health complications caused by mass accommodation in lodging houses with no sewage systems. Squalid conditions, the inability to sufficiently isolate the sick workers from the healthy, as well as the lack of basic standards of hygiene, according to officials, were in danger of causing epidemics of diseases in the spring months. During the year 1917, there were many cases of dysentery and various respiratory illnesses.[44] However, non-local workers housed in hastily built homes were not the only ones to live in unsuitable hygienic conditions. According to the official reports of the Prague military headquarters, the catastrophic standard of living affected almost all of the more than 40,000 workers employed in the Škoda factory.[45]

During the war years, a social problem with roots in the pre-war, liberal administration of the town escalated. Just like its wartime successor, it did not come up with any sort of comprehensive policy of urban development that would be able to prevent the destructive effects of the proletarianization of an ever-increasing number of factory workers. When the growth of the industrial labor force in Pilsen stopped increasing gradually, but between 1916 and 1917 jumped rapidly, working-class housing became out of control.[46]

These stages of industrial destitution demonstrated most obviously the complete social and functional segregation of wartime Pilsen, where the center of town was dominated by opulent middle-class buildings and apartment houses, while the outer circuits of the main industrial factories were littered with colonies of unprecedented poverty and social collapse. Rationalized housing was supposed to provide accommodation for workers after twelve-hour or longer shifts in the most economic way, but it became, along with workplaces and endless food lines, another space where the protest potential was emerging.

Workers of various nationalities, but with very similar biographies, met in hastily built workers' colonies. Economic migrants, arriving in Pilsen in the decades before World War I, were for the most part recruited from the ranks of the impoverished agricultural populations in the surrounding Czech-speaking villages.[47] One consequence of the massive wartime posting of more workers to the Pilsen Škoda branch in particular was, however, that the prewar, ethnically homogenous workers collective gained many German and Hungarian speaking colleagues, who, unless they came there directly from the front, had lived in a village for most of their lives.[48] Even though many of them had some experience with workers' organizations, for most of them being sent to Pilsen was a sudden and unexpected change. All of a sudden, they found themselves in one of the biggest industrial cities of the Habsburg Monarchy. Their integration into the existing workers' culture, which in Pilsen became deeply institutionalized during the prewar decades, was considerably problematic.[49]

The wartime culture of workers' colonies thus in no way functioned as a one-way accelerator of class consciousness under the conditions of an intensified struggle between work and capital, as historiography before 1989 postulated. In actual fact, it was drowning in its own inner dispute between established workers, who had settled in town before the war, and new arrivals, usually single and socially uprooted individuals, who had spent most of their lives in the countryside. "Class consciousness" was thus constantly confronted with the Christian morals of the countryside, emancipation efforts with the patriarchal perception of the social world and the workers' leaders' progressive language with the conservative worldview of the workers from the countryside.[50] However, the amount of time that the workers spent together led to the creation of a cohesive protesting subject, which constituted itself beyond language and social divisions. Thus, conscious and organized Czech workers could have been involved in the protests along with newly arrived veterans from the front, the sons of poor farmers as well as women, children or the elderly.[51]

Just like rations of work, rations of anger rose across-the-board, with no regard for gender, age or language. However, their specific materialization differed depending on individual biographies. Mass social protests represented the blending of pre-capitalist, country experiences with the capitalist formulas that workers adopted in their struggle against strict industrial discipline. The Pilsen railway workers' industrial protest—abandoning and then occupying their workplaces—could be compared to the enforcement of a pre-modern "moral economy." In the eyes of the workers from the countryside, the looted items were a rightly deserved reward for their work or merely goods, for which they paid a fair price. Originally, organized strikes in Slaný and Kladno could freely devolve into food theft and shop window smashing, which sometimes caused fatalities, without changing the composition of the protesters. The protesting subject, who had participated in many street riots in wartime Austria, eluded the officers and soldiers, who chased the looting bands through the city streets without success. It also escaped middle-class observers who, when they witnessed the spontaneous waves of violence and looting, were unable to make sense of it in the framework of their own economizing thought. The protesting subject of wartime Bohemia also proved elusive in much of the existing historiography, which tried to understand wartime mass protests through the perspective of the Austrian police and military forces. On the other hand, this subject also remained obscure to the working-class elite, which was trying in vain to integrate the newly arrived workers into advanced forms of organized workers' protests—mainly closed strikes behind factory walls, which were still the main form of organized working-class protests.

Although many strikes could freely turn into an uncontrolled wave of street violence, many of them maintained their prewar peaceful, organized form, limited to the factory and its occupants. Between 1914 and 1918 the collective refusal to work, isolated from the rest of the urban space, constituted the second central stage of the wartime protest politics, i.e., the moment when membership in the working class was formed and confirmed through collective protest.

The Spell of the Masses

The waves of mass street unrest, particularly during the last years of the war, turned the spatial organization of liberal Austrian towns upside down and represented a violent form of protest by a mixed group, in which workers were a significant, but not the only, part. With strikes it was often the opposite. Although some strikes freely turned into waves of street violence, many did not, and a large number maintained their pre-war character of organized working-class protests.

Historians have been traditionally interested in strikes. Communist historiography in particular saw them as proof of the growing class consciousness of the factory proletariat. Strikes were used as evidence of the transformation of the "class in itself" into a "class for itself"—a transformation of the collective of participants, who shared objective socioeconomic conditions, into a collective that was consciously fighting to change these conditions. Strikes thus provided a proof of the historic mission of the working class and of its inevitable "forward march" on its path toward absolute emancipation; proof of the working class's awareness of itself as the bearer of revolution and of the workers' purposeful undermining of the unjust bourgeois order.[52]

Pre-1989 historiography concentrated on the nature of the individual strikes' demands, and to a certain extent on the economic circumstances of their emergence. At the end of the socialist era, more differentiated works appeared, which tied the origin of strike movements to macroeconomic movements and focused on their influence on the material resources of the working class.[53] After 1989, Czech historiography to a certain extent abandoned the popular topic of workers' strikes altogether, and newer studies have been very rare.[54] Since the question of the origin of the strike collective was answered ideologically, i.e. by the process of the "awareness of the working class," pre-1989 historiography did not feel the need to concern itself further with the processes of the constitution of strike collectives. It also largely put aside their social composition, their symbols, and the demands that did not fit into the predefined templates of class awareness and struggle.

In order to depart from this ideological framework, which viewed strikes only as proof the working class's historically inevitable role as the avant-garde of communism, or from economic determinism, for which strike movements were born of economic forces beyond the control of the strike participants,[55] we must return to the strikers themselves and try to understand their own horizons and their influence on the origin and development of each strike. Within the context of the Bohemian lands and World War I, the widespread wave of strikes that swept through practically all of the industrial centers in the summer of 1917 is the best empirical example. The February overthrow of the Tsar in Russia was the spark that ignited the inner protest potential that had been building up in the Austro-Hungarian working class in the previous couple of years.[56] All of this resulted in the "hot summer" of 1917, when the majority of Bohemian towns were paralyzed by heretofore unseen strikes and unrests, marking a significant turning point in the history of wartime Austria and the first premonitions of the coming end of the monarchy.[57]

Although the strikes in the summer of 1917 had varying degrees of intensity, the largest ones were in the main industrial centers, especially Pilsen and Prague. At the same time, Pilsen and Prague represent the two different types of workers' strikes. A closer look at precisely these two cases can provide an insight into the various forms, as well as the limits, of the strikes, which the constrained wartime reality led to.

In Pilsen, the main epicenter of the organized strikes was the most important arms factory in the entire monarchy—Pilsen Škoda. There had been occasional refusals to work there already in 1916, but it was the strike at the end of June and the beginning of July 1917 that for the first time managed to paralyze the whole factory, and thus a significant portion of the Austrian arms production.[58] This mass refusal to work was the climax of the workers' fight against the military administrator of the factory, Kühn, and the factory director, Karabatschek, which had been raging since the spring. After the smaller unrests in March, the management of the factory was virtually flooded with workers' complaints about low wages that were well below inflation, but also about food, lodging, work safety and the treatment of workers by their superiors. The complaints dealt with various problems, from rotten food, the absence of ventilation, and disinfection in the latrines to rough treatment and beatings by the shift leaders.[59]

Since the army controlled the entire factory, working conditions were legally a military responsibility. The volume and contents of the complaints were so serious that the whole matter even reached the offices of the war ministry in Vienna. In May, the minister nominated a committee of generals, which was to get acquainted with the situation in Pilsen and propose a solution.[60]

However, not even this committee's investigation led to significant changes and in the second half of May, pressure in the Škoda factory intensified. The whole situation was further dynamized on May 25, 1917, when a series of explosions shook the nearby Bolevec ammunition plant, causing unprecedented devastation. One of the explosions of the stored ammunition unleashed a shock wave that traveled through the whole town, breaking windows in buildings more than 5 kilometers away. After the fires had been put out and the thick smoke had cleared, Austrian officials counted over 200 disfigured dead bodies. Despite strict safety measures, workers' leaders made radical speeches at the mass funerals of the explosion victims, and at the end of May and the beginning of June, Pilsen experienced the first organized Škoda workers' march, counting over 9,000 people.[61]

The whole situation escalated on June 26, 1917, when, instead of implementing the remedial measures that the committee of generals had promised a month earlier, the full militarization of the factory was announced.

Military supervision of the Škoda factory thus turned into full submission to military leadership. All of the workers, without exception, fell under military law and discipline.[62] Since almost none of the workers' earlier demands had been fulfilled, work ceased completely the next day. The first negotiating delegation that the workers sent out to talk with the factory management was arrested and imprisoned. The Škoda workers counteracted by refusing to obey calls to return to work, and around six o'clock in the afternoon they left the factory. Only a small number returned to work at seven o'clock in the evening for the night shift. Around eight o'clock that evening, several Škoda workers assembled in the garden of the *Peklo* social democratic house in Pilsen and debated their next steps. Their meeting was dispersed by an attending government official whom the workers respected, and they left peacefully.[63]

On the following day in the morning, roughly 15,000 workers met in front of the factory, but instead of going to work, the crowd set out to the nearby military training ground in "Borská pole," where the first mostly organized demonstration took place. Various speakers addressed the workers in Czech, German, Hungarian and Serbo-Croatian. The demonstration participants lined up in groups according to nationality, allowing for a small path in between, where a selected organizer gave out orders for the gathering to run smoothly. The speakers could easily identify the ethnic make-up of the listeners and each found his group, which he could address in the appropriate language.[64] All of the speakers called upon the participants to make a collective oath of solidarity and the crowd of workers subsequently swore it.[65]

Thusly organized camps took place every day and the workers were quickly joined by the social democratic Members of Parliament Franz Domes and Luděk Pik, who took over the negotiations with the military administration of the factory and regularly updated the workers at the demonstrations. Pik spoke Czech and Domes German, and their speeches were translated into other languages by selected workers, who stood by the particular national group.[66] Over the next couple of days, such gatherings took place regularly twice a day—at nine o'clock in the morning and two o'clock in the afternoon, and each meeting had the same agenda and course. Every time, more than 30,000 workers assembled, divided according to their language, and peacefully listened to the prepared rousing speeches. Then they all listened to the above-mentioned Members of Parliament, who delivered a progress report on the status of the negotiations. After that, the crowd dispersed peacefully. The police officers present had no reason to take action and the demonstrations ended calmly, usually with a call to continue the strike, maintain solidarity and share food so that none of the strikers would be forced to return to work for lack thereof.

In the remaining time, the workers returned to the factory as their work schedules dictated, but they continued to refuse to commence any kind of production.[67]

Due to the rising military presence at the "Borská pole," on the evening of June 19 a workers' meeting decided to move the morning demonstration to the "Na Zámečku" summer theater in the Pilsen neighborhood of Lochotín. Workers' guards were placed at the entrance of the theater to carefully check workers' membership cards and prevent any non-members from entering. At this assembly, Members of Parliament Pik and Domes announced that the military administrator of the factory, Kühn, was willing to accept several of the workers' demands and they advised the workers to return to work. In exchange, most of their demands would be fulfilled. Workers would in the future participate in the quality control of the food in the factory canteen, and their wages were to be significantly increased. Overtime and weekend pay was to be increased as well. The payment for individual tasks was to be posted on every machine in order to prevent arguments. All of those arrested at the beginning of the strike were promised immunity. The workday was to be shortened from twelve to nine hours and all of these concessions were to affect the whole staff, including the working women.[68]

On July 2, 1917, Škoda partially renewed production, but reaching maximum capacity would prove to be rather complicated. Hungarian and German workers in particular refused to return to work. Additionally, some workers left to stay with their families during the strike and since the press was banned from writing about the unrest, it was very difficult to inform them that the strike was over and that they needed to return to work.[69] Over a week passed before production in Pilsen was totally renewed and yet on July 9, 500 workers were still missing.[70]

The strike at the end of June and the beginning of July 1917 was the largest organized protest of industrial workers that the Bohemian lands had thus far experienced. According to official reports, only about 4,000 people continued to work in the Pilsen Škoda factory. With an overall number of more than 40,000 Škoda employees, this meant that the cessation of production was practically absolute.[71]

The entire strike, however, was characterized by an exceptional level of organization. All of the negotiations of the striking crowd, which numbered over 30,000 people, were peaceful. Demonstrations were perfectly spatially organized so that everyone could listen to the speeches in their native language, while the mass assemblies periodically met twice a day, always at the same time and same place, and never progressed to violent clashes with the patrolling police forces. At times the strike had almost a military style of discipline, which facilitated the repetitious movement of a large number of striking workers from the gates of the factory to the gathering place at the Borská pole or later at Lochotín, their subsequent splitting up into groups according to language, their calm attention to speeches and the exchange of food that allowed the poorest workers to participate in the strike, as they were the most affected by the closing of the factory canteen. As eyewitnesses noted, the number of workers who took part in the demonstrations and marches through town remained roughly the same throughout, despite the volatile weather, which oscillated between sweltering heat and powerful summer storms.[72]

Between the June 26 and July 2, 1917, an extraordinarily large and organized crowd of industrial workers, who managed to paralyze a key part of Austrian arms production, came together in Pilsen. Such a large and organized appearance by practically the whole staff of the factory, regardless of age, qualification or nationality, shocked not only the central military bodies in Vienna, but also the local administration, military and police force, which were otherwise quite well acquainted with the conditions of the local workers. The inability of the central government, the Prague governor's office, the Pilsen district governor, or the local armed forces to do anything about the strike for the entire week only underscored the government's surprise. The same can be said about the guarantee of immunity that the Austrian authorities provided to all of the strike participants, which was in essence the state's admission of defeat.

If we look more closely at the group of strikers and the course of the protest, we can see that the Pilsen strike differed in many respects from the established patterns of similar events that dominated working-class culture in the prewar era. Prior to the year 1914, strikes were generally conducted by a group of qualified men. Organized, collective appearances were the expression and confirmation of their masculine identity, structured around a dignified salary for their demanding qualified work, and solidifying their dominant position in the family. Workers' strikes generally reproduced the dominant ideal of separate spheres, within which public appearances were reserved mainly for men, while women's authority resided predominantly in the private sphere.[73]

However, the Pilsen strike did not adhere to this model at all. As is evident from the number of protesters and the reaction of the Austrian authorities, roughly 90 percent of the factory workers participated in the summer strikes in Pilsen and only around 4,000 people continued to work during the strike. Furthermore, in 1917 an unspecified number of prisoners of war from the Russian, Italian and Serbian fronts worked in the Škoda factory, for whom participating in the strike was out of the question. It is thus safe to say that the share of striking workers could have been very close to 100 percent.[74]

Not only the vast majority of qualified male workers, but also women and children participated in the mass work refusal in the Pilsen Škoda factory. The participation of women in such a large organized workers' protest, which in the prewar era was the clear manifestation of a masculine, i.e. rational, dignified and organized working-class culture was the first aspect that received significant attention from the Austrian administration as well as from the workers themselves. Gustav Habrman, a social democratic eyewitness to all of the assemblies and marches, observed with astonishment that during the summer protests in Pilsen, it was common to see "the figures of workers, men of the proletariat, hardened by strenuous work, living a hard life, together with the figures of women, their wives, the mothers of the next generation of workers."[75]

As is evident from the above quotation, in 1917 many workers' leaders were still thinking in categories that connected organized workers' protest with the masculine character of the protesting body. Women were thus not considered full-fledged members, but were seen, in a traditional manner, as wives and mothers. Nevertheless, even Habrman had to admit that women no longer played only supporting roles in the workers' strikes in the summer of 1917, but were, surprisingly to him, an equal part of the striking subject: "Men and women stood tightly packed next to one another. Everyone's heart beat rapidly. At times, a chill ran through their bodies, racked by hard work, poverty, suffering and hunger. One could sense that in their thoughts a question bounced around: What now? What will happen? … Pressure in the crowds rose every second. Shouts would be heard: So speak, then!"[76]

Habrman's observation of the workers' strikes, however, reveals another aspect besides the significant presence of women among the protesters. When the crowd of workers stopped to hear what was to come next, a wave of uneasiness swept over the strikers. What would happen next was unclear—would someone speak? And if so, who it would be and what he would say? Nevertheless, the participants were convinced that somebody should talk. Despite the high level of organization, the emergence of a coherent protesting subject was far from concluded; it was forming itself over the course of the protests. The protesting body was not merely the result of outside circumstances, such as the poor hygienic conditions in the workplace, low salaries, or the poor quality of the food. Although the first impulse to abandon work was certainly based on that, when it came time to decide on the speakers, leaders, and specific demands, the make-up of the crowd was still unfinished and could take any direction depending on what was said and who had said it.

Josef Hofmann, one of the main speakers who repeatedly spoke at the Borská pole at the end of June and the beginning of July, later admitted that the minute the whole crowd grew silent and anxiously waited for someone to give the assembly some content was, for him, the riskiest moment of the strike. He did not know until the last second whether he would have the courage to speak as he had agreed to with the other leaders of the workers' strike committee. As he himself admitted, he had never before spoken in front of such a crowd, and an oratorical fiasco on his part could have weakened the determination to strike in the more than 30,000-strong crowd, eager to hear someone's word.[77]

The final decision that the strike would continue in full force was reached during the speeches and later, when food was shared. It was at this moment that the strikers' indefinable sense of equality actually took form. Solidarity in terms of food distribution and listening to speeches together shaped the striking collective, so that when it returned to work, it was capable of hitherto unprecedented actions. The collective experience of the mass demonstration, shared so closely by so many participants that it was impossible to uphold the natural physical distances between human bodies and all of the participants, according to eyewitnesses, stood "packed tightly against one another,"[78] powered the creation of a collective that was absolutely equal. As Elias Canetti observed, only in such a dense mass

> may a person be free from the fear of being touched. It is the only situation in which this fear turns into its opposite. For this, a dense mass is needed, in which body is pressed upon body, dense even in terms of its psychological state, when we do not care who is crowding us. Once we have submitted to the crowd, we are not afraid of its touch. In the ideal crowd, all are equal. There are no differences, not even when it comes to gender. Whoever presses against us is the same as us. We can feel him just as we can feel ourselves. Then everything happens as if inside a single body. … In the throng, there is practically no space between us, body is pressed against body, one is as close to another as to oneself … [in this moment] nobody is better than another.[79]

Physical proximity, caused by the immense density of bodies during the workers' gathering, which under any other circumstances would have been rather uncomfortable, provided the participants with a framework for identification with the entire strike movement, and also erased any differences between the strikers. Everyone could feel as though they were part of a "mass body." The emphasis on the close connection of the bodies of the protesting workers was thus paramount not only during the first day of the strike, but also in its later stages. Whenever possible, the striking workers created very tight processions, in which their bodies were pressed against each other. These processions thus provided participants with a feeling of personal safety and collective strength. For many workers, this feeling of collective togetherness would never be repeated and was etched in their memories as the strongest part of the whole strike movement. "We marched in rows of twenty as a mighty river, crammed from wall to wall, hand in hand. Pilsen was overwhelmed by the sense of the united will of the Škoda workers,"[80] remembered one of the participants of the workers' marches that flooded Pilsen daily at the end of June and the beginning of July 1917.

This was a relatively new moment in the tradition of working-class mass marches. Despite the comparatively similar number of participants, the prewar working-class demonstrations generally did not cross the threshold of basic physical comfort. In fact, participants often anxiously took care to maintain physical autonomy, and the protest collectives were not strengthened by the intense experience of mass physical togetherness. Usually, only men would take part in protest marches and women would join in only after they were finished, so that they could enjoy the following evening party with their partners. When, for example, one of the major nineteenth-century Czech journalists, Jan Neruda, described the appearance of a workers' crowd that marched through Prague on May 1, 1890, it did not escape his notice that "mobs are coming. Not dense, purposely thin, thus infinite. … But in the streets, the flow went on unabated. Not during the day and not in the evening. On the contrary, there were more and more people hour after hour. After that, workers walked proudly with their wives by their right side."[81]

In Pilsen in 1917, however, the striking subject was shaped by the intense feeling of collective togetherness and solidarity that blurred the distinctions not only between age and qualification, but also between the sexes, and began to exist and act as a coherent agent, consisting of more than 30,000 bodies. Such an agent was then able to change the world around it in a way that was previously unthinkable. After the strikers returned to the Škoda factory, nobody resumed work. The vast production halls of the Pilsen Škoda factory, normally dominated by the loud omnipresent noise of machines, were all of a sudden transformed into a place dominated by a persistent silence; motionless tranquility replaced the constant hustle and bustle of bodies and machines.[82] The vast production spaces thus shed their everyday character and transformed into a kind of sacred place—characteristically wide and silent, its disposition disciplining all who moved inside it.[83] A former place of ever-present industrial rationalization suddenly became a symbol of the total, albeit fleeting, dominance of the strikers.

This complete reversal of the significance of the workspace at the same time opened new horizons of thought for all of the participants. Working men and women were no longer the factory management's mere rationalized objects or the targets of physical violence at the hands of employers, but they became an entirely new entity—a class in the true meaning of the word, which was not only capable of a one-time collective act, but with its own expectations and notion of the future, very different from those of the majority. Previously powerless workers, deprived of their rights and in many cases of their own dignity, could now be a part of an all-mighty project that could completely silence, and thus symbolically re-code, immense industrial spaces—a project that enabled them to see their current suffering as something temporary that could in the future be changed to ubiquitous silence and peace, which also symbolized the post-war restoration.[84]

The fact that the striking workers were able to totally redefine the factory space and maintain its new meaning without any problems even throughout the subsequent days shows not only the incompetence of the Pilsen military garrison and police, but also the strength of the unified strikers. This unity could not draw on any pre-war roots, for the old workers' organizations in Pilsen were disbanded due to the war draft and a great number of newly conscripted workers came from every corner of the Monarchy. An extraordinarily coherent and active strikers' group was nevertheless created during the strike itself, influenced by their shared material conditions, but especially by the protest's particular form and organization, which had the considerable potential to create an autonomous collective subject out of an amorphous mass of men, women, and children of various ages, professions, income, and spoken languages.

The presence of women and adolescents in the striking group symbolized the radical departure from the prewar strike tradition, when women in most group protests were marginalized and generally perceived as irrational and irresponsible elements.[85] The presence of every profession, divided into very lucidly defined income groups within the wage system of the Pilsen Škoda factory, created a considerably heterogeneous group in terms of salary. Again, this symbolized the bridging of all economic dividing lines.[86] While in the prewar era the core group of strikers mostly came from qualified working-class professions, whose self-perceptions were tied to their ability to provide valued work for sufficient pay, wage levels or qualifications did not play a role in the Pilsen strike in the summer of 1917. Finally, the presence of all of the nationalities represented in the Škoda factory symbolized the crossing of the dividing lines of language and nationality. The organization of the demonstrations, in which each nationality had a predetermined place and where the speeches were either directly spoken in each necessary language or immediately translated, allowed the strikers to join the group regardless of nationality or language. The similarly shared experience from the intimate crush of thousands of human bodies, or from the opening up of new horizons of imagination with the symbolic re-marking of the industrial spaces, did not take the languages of the individual participants into consideration and thus crossed a potential ethnic dividing line.

The significance of such an experience for the organized workers' protest becomes even clearer if we look at the second stage of the mass strikes in the summer of 1917, which took place in the Bohemian capital of Prague. As the largest city in the Bohemian lands, it, like Pilsen, went through significant demographic changes during the war years. The city of roughly 300,000 inhabitants rapidly expanded with the number of industrial workers, who constituted more than a third of the population in several Prague neighborhoods. This constantly expanding working-class environment generated, especially in the last two years of the war, an increasing number of various protests and became the main source of worry for the municipal and state administration.[87] Like the rest of the Bohemian lands, Prague witnessed a significant turning point during the summer of 1917, when the first large wave of unrests and strikes took place, which foreshadowed the escalating waves of protests in the following year, eventually leading to the total collapse of the Austrian Empire.

The first signs of the summer unrest began to appear in the middle of May 1917, when the Czech governor's office for the first time nervously informed its subordinate offices that "among the workers of the wartime industry something is boiling, manifesting itself in numerous, albeit brief, refusals to work, increasing demands for higher wages and complaints of bad supplies."[88] As the governor's office itself noted, the growing number of displays of unhappiness was connected with the inner shifts within the working class itself. New, radical workers' secretaries came to the forefront of various protest acts, gaining prominence in individual factories at the expense of the current, mostly social democratic workers' leaders. Organized workers' events thus slowly evaded the traditional domination of the Social Democratic Party that shaped almost all of the organized industrial protests of the prewar era.[89]

The smoldering dissatisfaction first rose to the surface during the final days of May, when on May 30 and 31, 1917, many workers in the main industrial factories in Prague and its surroundings left their work. Operations were partly stopped, for example, in the Kolben electrotechnical company in the Prague neighborhood of Vysočany,[90] the Ringhoffer train car factory in Smíchov, in the František Křižík electrotechnical factories in Karlín and in many smaller factories in Holešovice, Libeň, Bubny, and other Prague working-class neighborhoods.[91] However, none of these work refusals were spontaneous and individual acts. Workers from different factories closely coordinated amongst each other, and during the strike itself the factories sent their representatives to other participating production sites, where these emissaries spoke at workers' meetings, assuring the employees of the respective factory that they were not alone in the strike and that it was worth continuing. The 30-year-old assembler Josef Donát from the Křižík plant and the three-years-younger lathe operators Jaroslav Šulc and Adolf Kratochvíl, for example, constantly shuttled between all the factories that formed the axis of the Karlín strikes during the unrests. The 20-year-old plumber Josef Krejčí from the Ringhoffer train car factory in Smíchov took care of communications between the local factories and the mobilization of the local protesters by repeatedly going around the striking factories. He always introduced himself as the representative of the metal workers' union, made a speech to the assembled strikers and called upon them to leave their factories and participate in a demonstration march through the streets of the city.[92]

In the end, the march did not take place, and the May Prague strike limited itself to repeated, demonstrative suspensions of work and several workers' assemblies that articulated some of the strikers' demands. Nevertheless, the partial refusal to work in the central factories of the Prague machine and electrotechnical industry disturbed the local police and military officers. The larger Prague companies in general were subject to the law on state-protected enterprises and were therefore, like the Pilsen Škoda factory, vitally important for the entire war effort. The Prague military headquarters thus paid close attention to this first wave of strikes and in the beginning of July 1917, identified in the Ringhoffer train car factory, for example, a total of eighteen main strike culprits, who were immediately punished and sent away from Prague.[93]

However, the mass strikes in Prague were not stopped by this intervention. Occasional refusals to work continued to appear in many Prague factories, and at the end of July and the beginning of August these strikes went beyond the electrotechnical and mechanical factories and spread to the railways and to the Prague municipal transportation system, particularly among the tram staff. Unrest culminated between August 6 and 9, 1917, when the largest organized strike in Prague took place. It began on August 6 in the Prague metalworks factories, when the hastily formed workers' assembly decided to send its representatives to the Bohemian governor's office and to the government in Vienna.[94]

The next day, the strike spread to several electrotechnical factories, the Prague municipal transportation system, and the Prague railway network.[95] The danger of combining these strikes into one unified movement that could culminate in a mass uprising, according to the estimates of the Prague police bureau, grew ever greater. Since the organized refusal to work spread to the Prague railway and tram depots and stops during the first days of August, there was an acute risk of the total paralysis of not only the municipal but also domestic rail transportation, which would have far-reaching consequences for the entire wartime provisioning.[96] But in end, the Austrian military and police authorities' doubts proved groundless. There were no mass protests that would imminently threaten the overall stability of the hinterland, and the strikers limited themselves to a few protest meetings during July and August, where they repeatedly called for higher wages, shorter workdays, etc. Only very few of their demands were truly fulfilled.[97]

The Prague strikes from the summer of 1917 never reached the same level of domination as in Pilsen. Most of the workers' demands were not fulfilled and the strikers were thus not successful. Quite to the contrary—when the Prague military headquarters continuously assessed the turbulent situation in the Bohemian lands during the summer of 1917, it used the development of the strikes in Prague as a textbook example of an unsuccessful wave of strikes and the collapse of organized workers' protests.[98] The local authorities' explanation as to why the Prague collective protests were doomed to fail can provide a deeper view into the composition and proceedings of the strike collective.

As the Prague police bureau noted shortly after the end of the largest wave of strikes between August 6 and 9, the whole strike movement "suffered from a lack of inner solidarity among the workers, which made itself visible especially in the large number of strikebreakers and in the deep conflicts between the male and female staff.... The female tram staff announced during the meeting that in the matter of the demands for higher wages it would be satisfied with tips."[99] The Prague military headquarters made a similar judgment based on its information, seeing many dividing lines among the protesters, rather than a unified collective. However, it perceived the main division to be not along the lines of gender, but along the lines of generation and nationality: "It has become clear that it is mainly the young and rash elements that are contributing to the terror upon the other workers, so it will be necessary to eliminate them from their current workplace as soon as possible.... It is vitally important to send these people away to work in factories, where a majority of workers speak other languages. In this way they will be rendered harmless."[100]

This judgment by the Austrian state bodies—seeing the Prague striking collective as a conglomerate of several mutually incompatible groups, often with contradictory interests—was not pointless. If we look more closely at the course of the summer strikes in Prague and at the strikers themselves, we can see that the Prague "hot summer of 1917" was, in fact, very different from the Pilsen one in many aspects. Already at the beginning of the Prague unrests in the last days of May, it became apparent that the striking group would not be able to integrate the entire Prague working class of a certain sector, as was the case in the Pilsen Škoda ammunition and arms plants.

The May Prague strike was not a collective mass protest, but rather a project shaped by young male workers, who organized it, kept it going and gave it its specific course and form. Almost all of the workers' activists identified by Austrian authorities were younger than 30 and had qualified professions. The above-mentioned 27-year-old lathe operators Šulc and Kratochvíl from the Křižík factories and the 22-year-old plumber Krejčí from the Ringhoffer train car factory, like the vast majority of the other main figures of the Prague strikes, belonged to the youngest generation of Czech-speaking qualified workers. While there were many women among the strikers and even among the arrested and interrogated in Pilsen, not a single woman was arrested in Prague.[101]

The composition of the striking collective similarly differed with regard to age. In Pilsen, factory operations stopped completely and older as well as younger workers participated in the strikes. In contrast, the summer strike in Prague was mostly the domain of workers, whose age rarely exceeded 30 years. Despite the fact that the Pilsen as well as the Prague strikes were characterized by an emphasis on organization, peace and the careful voicing of the workers' demands, the protests were disrupted by occasional violent incidents more often in Prague. In Pilsen, the strike was peaceful from beginning to end, and even the otherwise strict police force did not find a reason to intervene against the disturbance of public order. In Prague, on the contrary, factory equipment was repeatedly damaged, doors were broken down, and violence was perpetrated upon workers who refused to participate in the strike.[102] The greater number of these incidents attests to the disunity of the Prague workers, causing severe internal tensions within the respective work force.

Next to gender and age, the third main dividing line was the strikers' ethnic composition. While in Pilsen the mass strikes were always accompanied by efforts to include all of the participating languages and nationalities, evinced by the spatial organization of the strike assemblies, their progress, etc., in Prague the strike was limited solely to Czech-speaking workers. This was reflected in some of the workers' demands, such as the call to release the Czech national-socialist representative Václav Klofáč from Austrian prison.[103] The Prague military headquarters could thus break up the strike in many factories fairly easily simply by mixing the nationalities in the working groups, because, just like in Pilsen, there was a sufficient number of non-Czech speaking workers in Prague.[104] The disposition and space of both industrial cities likely played a role. In Pilsen, most of the industries were concentrated in one neighborhood, but in

Prague the factories were strewn across various neighborhoods, which could make communication between the factories significantly more difficult. All of the internal dividing lines in the Prague working class were not overcome like they were in Pilsen, resulting in the overall ineffectiveness of the strike, which eventually faded to nothing.

An episode that took place on the afternoon of May 31, 1917 before the gates of the Kutzer Company in the Prague working-class district of Smíchov is illustrative. On the second day of the May protests, the company workers gathered to hear the strike emissary, who was sent from the Ringhoffer train car factory and who was to explain the purpose, course, and goal of the strike. The emissary was the previously mentioned 27-year-old plumber Josef Krejčí, who had been going back and forth between the factories in Smíchov since the day before and was trying to mobilize the workers for the ongoing strike. Krejčí came before the assembled workers, introduced himself as a representative of the Czech Metal Workers' Union, and appealed to them in his name to line up under its leadership and head out on a march through the city. When the attending police officer asked him to prove his relationship with the previously mentioned union, he was not able to answer. After several questions, it became apparent that Krejčí was just pretending to be affiliated with the largest Czech union in order to give his words more weight when trying to convince the waiting workers.[105] His desperate attempt to mobilize the assembled workers behind him, even at the cost of providing false information, only underscored the poor organization of the strike and the desperate desire of young male qualified leaders to take control of its leadership at all costs, be it through physical violence or simple lies.

While in Pilsen the strikers stood together in impressive formations of up to 30,000 people over the course of several days, in Prague the workers' assemblies only rarely numbered more than 1,000.[106] Thus, in the case of the Prague strike, it is hard to imagine the formative experience of mutual solidarity, the feeling of strength, and the safety that ensued from being part of an enormous mass, which many participants of the Pilsen unrests attested to. Quite to the contrary, young, qualified Czech-speaking male workers, who often defied the traditional social democratic dominance of such events and were more in line with the radical National Socialist party, dominated from the beginning and prevented the emergence and further strengthening of a coherent protesting mass.[107] The violent behavior inherent in the organization of the strike as well as the exclusion of women and non-Czech speaking workers caused it to fade away without considerable success.

Pilsen and Prague thus represent two opposing types of organized strike movements that transformed wartime Austria since the summer of 1917. Rising anger in reaction to the changes in the worlds of material consumption, physical work and gender relationships, could, alongside popular uprisings, materialize in various forms and various intensities in accordance with the striking groups' specific performative practices. The Pilsen strikers' experience of close physical proximity and merging with a large, unified body created a sense of unbreakable solidarity, with no consideration for age, qualification, gender, or nationality. This solidarity formed the background of a mass resolution to strike and was the source of some success for the strikers. In Prague, however, such a formative experience could not be possible due to the exclusive dominance of young, qualified Czech-speaking men; the strike was therefore destined to fail. The Pilsen type of strike was more flexible and able to adapt to the changed wartime environment, and to integrate groups that would never have been part of it before the war. It could therefore achieve the fulfillment of most of its demands. The Prague strikers represented the opposite type, which considered the qualified male worker to be the only subject of working-class interests, and thus remained embedded in the prewar tradition of organized workers' protests. These, however, no longer corresponded to the changed wartime working-class environment and thus could not be successful.

Notes

1. Johannes Urzidil, "Karl Weissenstein," in Johannes Urzidil, *Pražský triptych* (Prague, 1997), 150.
2. Šedivý, *Češi, české země a velká válka 1914–1918,* 320–21.
3. Nedvěd, *Jak to bylo na českém západě 1914–1918,* 140.
4. Healy, *Vienna and the Fall of the Habsburg Empire,* 81–82.
5. Protocol from the ministerial meeting on the possibility of declaring martial law during strike incidents on August 23, 1916, in Rudolf Neck, ed., *Arbeiterschaft und Staat im Ersten Weltkrieg,* document no. 77, 94–97.
6. Šedivý, *Češi, české země a velká válka 1914–1918,* 319.
7. Peter Heumos, "Kartoffeln her oder es gibt eine Revolution. Hungerkrawalle, Streiks und Massenproteste in den böhmischen Ländern 1914–1918," *Slezský sborník* 97, no. 2 (1999): 81–104.
8. The subsequent course of the protests is described in great detail in the report of the Pilsen district office. See District office in Pilsen to the police directorate in Prague on August 14, 1917. Report on hunger demonstrations in Pilsen on August 13, accompanied by the looting of shops, and on the declaration of martial law over Pilsen, Škvrňany a Lobzy, in *Sborník dokumentů k vnitřnímu vývoji českých zemí za 1. světové války,* vol. IV, document no. 63, 150–53.
9. Ibid.
10. Nedvěd, *Jak to bylo na českém západě 1914–1918,* 141.

11. Disctrict office in Pilsen to the police directorate in Prague on August 14, 1917. Report on hunger demonstrations in Pilsen on August 13, accompanied by the looting of shops, and on the declaration of martial law over Pilsen, Škvrňany a Lobzy, in *Sborník dokumentů k vnitřnímu vývoji českých zemí za 1. světové války*, vol. IV, document no. 63, 150.
12. Heumos, "Kartoffeln her oder es gibt eine Revolution," 85.
13. Nedvěd, *Jak to bylo na českém západě 1914–1918*, 142.
14. See, for example, Josef Kolejka and Václav Peša, "V boji proti válce i proti monarchii 1917–1918," in *Z dějin dělnického hnutí na Brněnsku. Od nejstarších počátků do založení KSČ*, ed. Josef Kolejka (Brno, 1956), 144–55; Josef Kolejka, *Revoluční dělnické hnutí na Moravě a ve Slezsku 1917–1921* (Prague, 1957), 43–59.
15. Heumos, "Kartoffeln her oder es gibt eine Revolution," 81–82.
16. Report of the Czech governor's office to the Ministry of the Interior on the situation and the mood in the Bohemian lands on August 3, 1917, in Neck, *Arbeiterschaft und Staat im Ersten Weltkrieg*, document no. 256, 31–32; and the report of the Moravian governor's office to the Ministry of the Interior on the situation and mood in Moravia from August 4, 1917, in ibid., document no. 261, 36–38.
17. See, for example, Kárný et al., *Sto let Kladenských železáren*, 273–82.
18. Heumos, "Kartoffeln her oder es gibt eine Revolution," 81–104; Berthold Unfried, *Arbeiterprotest und Arbeiterbewegung in Österreich während des Ersten Weltkrieges* (Vienna, 1990).
19. Šedivý, *Češi, české země a velká válka 1914–1918*, 318–26.
20. Ibid., 320.
21. Ibid., 321.
22. Libuše Otáhalová, ed., *Souhrnná hlášení presidia pražského místodržitelství o protistátní, protirakouské a protiválečné činnosti v Čechách 1915–1918* (Prague, 1957), especially 191–410.
23. CMAP, 9. Sborové velitelství (1883–1919), karton 115, č. res. 23713, Hlášení z Teplic 9. sborovému velitelství v Litoměřicích z 31. května 1917.
24. District office in Pilsen to the police directorate in Prague on August 14, 1917. Report on hunger demonstrations in Pilsen on August 13 accompanied by the looting of shops, and on the declaration of martial law over Pilsen, Škvrňany a Lobzy, in *Sborník dokumentů k vnitřnímu vývoji českých zemí za 1. světové války*, vol. IV, document no. 63, 150.
25. Plaschka, Hasselsteiner and Suppan, *Innere Front*, 202.
26. Telegram of the Vítkovice railway directorate to the ministry of the interior from July 3, 1917, in *Arbeiterschaft und Staat im Ersten Weltkrieg*, vol. 1, document no. 245, 17.
27. Martin Jemelka, ed., *Lidé z kolonií vyprávějí své dějiny* (Ostrava, 2009), 329.
28. Plaschka, Hasselsteiner and Suppan, *Innere Front*, 257.
29. Report of the Czech governor's office to the police director in Prague and presented to several disctrict governor's offices in the Bohemian lands from February 1, 1918, in *Sborník dokumentů k vnitřnímu vývoji českých zemí za 1. světové války*, eds. Jaroslav Vrbata and Eva Drašarová (Prague, 1997), vol. 5, document no. 16, 71–74.
30. Maderthaner and Musner, *Unruly Masses*, 3–7.

31. See the influential postcolonial studies text Dipesh Chakrabarty, *Provincializing Europe. Postcolonial Thought and Historical Difference* (Princeton, NJ, 2000), especially 72–96.
32. Edward P. Thompson, *The Making of the English Working Class* (London, 1961); Edward P. Thompson, "The Moral Economy of the English Crowd in the Eighteenth Century," *Past and Present* 50, no. 1 (1971): 76–136; James C. Cott, *The Moral Economy of the Peasant: Rebellion and Subsistence in Southeastern Asia* (New Haven and London, 1976).
33. CMAP, 9. Sborové velitelství, karton 149, sign. 72-38/9, Hlášení okresního hejtmanství Mladá Boleslav 9. Sborovému velitelství z 20. června 1918.
34. CMAP, 9. Sborové velitelství, karton 149, sign. 72-38/9, Hlášení vedení podniku Laurin a Klement 9. sborovému velitelství z 30. března 1918.
35. Military command in Prague to the imperial military office in Baden. Report on supply conditions, opinions and the economic situation of the working class in the Bohemian lands from March 14, 1917, in *Sborník dokumentů k vnitřnímu vývoji českých zemí za 1. světové války*, vol. IV, document no. 13, 55.
36. Habrman, *Mé vzpomínky z války*, 24.
37. Leonardo Benevolo, *Die Stadt in der Europäischen Geschichte* (München, 1993), 192–93.
38. Nedvěd, *Jak to bylo na českém západě 1914–1918*, 141.
39. Maderthaner and Musner, *Unruly Masses*, 28–29.
40. Ibid., 108–09.
41. Fasora, *Dělník a měšťan*, 199–210.
42. Cf. Hanns Haas and Hannes Stekl, ed., *Bürgerliche Selbstdarstellung: Städtebau, Architektur, Denkmäler* (Vienna, 1995); Michaela Marek, "Monumentalbauten und Städtebau als Spiegel des gesellschaftlcihen Wandels in der 2. Hälfte des 19. Jahrhunderts," in *Böhmen im 19. Jahrhundert*, ed. Ferdinand Seibt (Frankfurt am Main, 1995), 149–223.
43. Lieske, *Arbeiterkultur und bürgerliche Kultur*, 54–57.
44. Janáček, *Největší zbrojovka monarchie*, 380–81.
45. Transcript of the dispatch of the Prague military command to the ministry of war in Vienna from July 17, in Nedvěd, *Jak to bylo na českém západě 1914–1918*, 158.
46. Janáček, *Největší zbrojovka monarchie*, 378–82.
47. Lieske, *Arbeiterkultur und bürgerliche Kultur*, 57.
48. Janáček, *Největší zbrojovka monarchie*, 378–80.
49. Lieske, *Arbeiterkultur und bürgerliche Kultur*, 95–109.
50. See, most recently, for the case of Belgium, Antoon Vrints, "Beyond Victimization: Contentious Food Politics in Belgium during World War I, " *European History Quarterly* 45, No. 1 (2015): 83–107.
51. Habrman, *Mé vzpomínky z války*, 83–90.
52. For the most significant works in this respect see Zdeněk Šolle, *Dělnické stávky v Čechách v druhé polovině 19. století* (Prague, 1960); Jaroslav Purš, *Postavení dělnické třídy a stávkové boje v českých zemích v období průmyslové revoluce* (Prague, 1964); Josef Polišenský, Karel Novotný and Věra Vomáčková, *Boj dělníků na stavbách našich prvních železnic* (Prague, 1956); Zdeněk Šolle, "K počátkům

dělnického hnutí v Praze," *Československý časopis historický* 5 (1957): 664–87; Andělín Grobelný and Bohumil Sobotík, *Dělnické hnutí na Ostravsku* (Ostrava, 1957); Josef Kolejka, ed., *Z dějin dělnického hnutí na Brněnsku. Od nejstarších počátků do založení KSČ* (Brno, 1956).

53. Jana Machačová, *Mzdové a stávkové hnutí dělnictva v průmyslových oblastech českých zemí 1848–1914* (Opava, 1990). For an overview of the pre-November research on working-class strikes, see Jana Machačová, "Výzkum stávek z 19. a 20. století v československé historiografii. Přístupy českých a slovenských historiků." *Studie k sociálním dějinám* 2 (1998): 139–51; Stanislav Knob, "Stávkové hnutí v československé i zahraniční historiografii. Srovnání a výhledy," in *Problematika dělnictva v 19. a 20. století*, ed. Stanislav Knob and Tomáš Rucki (Ostrava, 2011), 65–72.
54. Knob, "Stávkové hnutí v československé i zahraniční historiografii," 69.
55. See, for example, Leopold D. Haimson and Charles Tilly, ed., *Strikes, Wars, and Revolutions in International Perspective: Strike Waves in the Late Nineteenth and Early Twentieth Century* (Cambridge and New York, 2002).
56. Galandauer, *Ohlas Velké říjnové socialistické revoluce a české země.*
57. Šedivý, *Češi, české země a velká válka 1914–1918*, 318–26; Heumos, "Kartoffeln her oder es gibt eine Revolution," 89.
58. District governor's office in Pilsen to the presidium of the Czech governor's office in Prague. Report on the declaration of the militarization of the Škoda factories in Plzň and the reaction of their employees from July 30, 1917, in *Sborník dokumentů k vnitřnímu vývoji českých zemí za 1. světové války*, vol. IV, document no. 45, 112–19.
59. Nedvěd, *Jak to bylo na českém západě 1914–1918*, 185–95.
60. Janáček, *Největší zbrojovka monarchie*, 390.
61. Ibid., 392–95; Antonín Nedvěd, *Katastrofa v Bolevci 25. května 1917* (Pilsen, 1924).
62. *Večerník Práva lidu* 216 (August 8, 1917): 3.
63. District governor's office in Pilsen to the presidium of the Czech governor's office in Prague. Report on the declaration of the militarization of the Škoda factories Pilsen and the reaction of their employees from July 30, 1917, in *Sborník dokumentů k vnitřnímu vývoji českých zemí za 1. světové války*, vol. IV, document no. 45, 112.
64. Habrman, *Mé vzpomínky z války*, 130–31.
65. District governor's office in Pilsen to the presidium of the Czech governor's office in Prague. Report on the declaration of the militarization of the Škoda factories in Pilsen and the reaction of their employees from July 30, 1917, in *Sborník dokumentů k vnitřnímu vývoji českých zemí za 1. světové války*, vol. IV, document no. 45, 112–13.
66. Ibid.
67. Report of the Prague military procurator to the military command in Prague from July 16, 1917, in Nedvěd, *Jak to bylo na českém západě 1914–1918*, 210–14.
68. Habrman, *Mé vzpomínky z války*, 138–39.
69. Report of the Prague military procurator to the military command in Prague from July 16, 1917, in Nedvěd, *Jak to bylo na českém západě 1914–1918*, 219.

70. Ibid., 220.
71. District governor's office in Pilsen to the presidium of the Czech governor's office in Prague. Report on the declaration of the militarization of the Škoda factories in Pilsen and the reaction of their employees from July, 30, 1917, in *Sborník dokumentů k vnitřnímu vývoji českých zemí za 1. světové války,* vol. IV, document no. 45, 115–16.
72. Habrman, *Mé vzpomínky z války,* 135–36.
73. Mareš and Strobach, "Třída dělníků i žen?," 34–68.
74. Janáček, *Největší zbrojovka monarchie,* 376–82.
75. Habrman, *Mé vzpomínky z války,* 113.
76. Ibid., 110–11.
77. Nedvěd, *Jak to bylo na českém západě 1914–1918,* 199.
78. Ibid.
79. Elias Canetti, *Masa a moc* (Prague, 1994), 10–13.
80. Quoted in Čepelák et al., *Dějiny Plzně,* 232.
81. Jan Neruda, *Fejetony* (Prague, 2011), 120.
82. District governor's office in Pilsen to the presidium of the Czech governor's office in Prague. Report on the declaration of the militarization of the Škoda factories in Pilsen and the reaction of their employees from July 30, 1917, in *Sborník dokumentů k vnitřnímu vývoji českých zemí za 1. světové války,* vol. IV, document no. 45, 112–19.
83. This passage is inspired by the analysis of the mass protests in pre-war Vienna by Wolfgang Maderthaner and Lutz Musner. See Maderthaner and Musner, *Unruly Masses,* 118–20.
84. Ibid., 141–42.
85. See, for example, Václav Holek, *Paměti: společná cesta české a německé sociální demokracie koncem devatenáctého století* (Prague, 2011), 159–60.
86. Miroslav Eisenhammer, "Mzdová hladina zaměstnanců Škodových závodů v Plzni v době první světové války," *Historie a vojenství* 4 (2000): 766–79.
87. ACCP, Referát XIV. Zásobování, inv. no. 31, Protokoly aprovisační komise za rok 1916, protokol z 29. srpna 1916.
88. Presidium of the Czech governor's office to subbordinate political offices on May 14, 1917, in *Sborník dokumentů k vnitřnímu vývoji českých zemí za 1. světové války,* vol. IV, document no. 30, 84.
89. Ibid.
90. For the development of the Kolben and Křižík factories, cf. Marcela C. Efmertová, *Elektrotechnika v českých zemích a v Československu do poloviny 20. století* (Prague, 1999), 62–67; Jana Geršlová, "Elektroindustrie in den böhmischen Ländern: Emil Kolben und die Entwicklung der Firma ČKD Prag," in *Business History. Wissenschaftliche Entwicklungstrends und Studien aus Zentraleuropa,* eds. Alice Teichová, Herbert Matis, and Andreas Resch (Vienna, 1999), 197–203.
91. Prague police directorate to the military command in Prague. Report of the Prague police directorate to the military command in Prague on the course of working-class demonstrations from May 30 to 31, 1917, from July 2, 1917, in *Sborník dokumentů k vnitřnímu vývoji českých zemí za 1. světové války,* vol. IV, document no. 37, 97–98.

92. Ibid.
93. Order of the Prague military command to investigate 18 workers of the Ringhoffer factories in Prague under suspicion of being the main instigators of the May working class unrests, July 5, 1917, in *Sborník dokumentů k vnitřnímu vývoji českých zemí za 1. světové války,* vol. IV, document no. 48, 121–22.
94. *Právo lidu* 215 (August 7, 1917): 5.
95. LUAP, Staré odborové spolky, sign. 3180/246A, Zápisník schůzí dělnického výboru Prague—Bubny, Zápis z 10. srpna 1917.
96. Report of the Prague police directorate to the Czech governor's office from August 8, 1917, in *Sborník dokumentů k vnitřnímu vývoji českých zemí za 1. světové války,* vol. IV, document no. 61, 146–48.
97. Report of the Prague police directorate to the Czech governor's office from August 10, 1917, in ibid., document no. 62, 148.
98. Report of the military command in Prague to the Czech governor's office from July 14, 1917, in ibid., document no. 54, 137–38.
99. Report of the police directorate in Prague to the presidium of the Czech governor's office from August 10, 1917, in ibid., document no. 62, 150.
100. Report of the military command in Prague to the Czech governor's office from July 14, 1917, in ibid., document no. 54, 138.
101. Nedvěd, *Jak to bylo na českém západě 1914–1918,* 220.
102. Order of the Prague military command to investigate 18 workers of the Ringhoffer factories in Prague under suspicion of being the main instigators of the May working-class unrests, July 5, 1917, in *Sborník dokumentů k vnitřnímu vývoji českých zemí za 1. světové války,* vol. IV, document no. 48, 121–22.
103. Police directorate in Prague to the presidium of the Czech governor's office, September 29, 1917, in ibid., document no. 76, 182–89.
104. For the nationalization of the working class environment before and during World War I, see the most recent Jakub Beneš, "Czech Social Democracy, František Soukup, and the Habsburg Austrian Suffrage Campaign 1897–1907: Toward a New Understanding of Nationalism in the Workers' Movements of East Central Europe," *Střed/Centre* 2 (2012): 9–33; Václav Houfek, "Nacionalizace společnosti a dělnictvo na Ústecku do roku 1918," in *Nacionalizace společnosti v Čechách 1848–1914,* eds. Kristina Kaiserová and Jiří Rak (Ústí nad Labem, 2008), 281–315.
105. Prague police directorate to the military command in Prague. Report of the Prague police directorate to the military command in Prague on the course of working-class demonstrations from May 30 to 31, 1917, from June 2, 1917, in *Sborník dokumentů k vnitřnímu vývoji českých zemí za 1. světové války,* vol. IV, document no. 37, 97–98.
106. Prague police directorate to the military command in Prague. Report of the Prague police directorate to the military command in Prague on the course of working-class demonstrations from May 30 to 31, 1917, from June 2, 1917, in ibid., document no. 76, 182–89. Further see Galandauer, *Bohumír Šmeral 1914–1941,* 128–29.
107. Kárník, *Habsburk, Masaryk, či Šmeral,* 163–65.

Conclusion

The Habsburg Monarchy's efforts in fighting a total war ushered in many societal changes in the last four years of its existence, greatly influencing the shape and agenda of the working class, which was able to continue in its prewar tradition only in limited fashion. The crippling of the Empire's social and political life, together with massive relocations to the front, practically destroyed prewar working-class culture. The wartime working class thus had to reconstitute itself in a new form and under new conditions.

The preceding pages analyzed the specific ways in which prewar workers' organizations were disbanded, as well as moments when the working class reappeared as a collective subject. Based on the examples of four spheres in which the reconstitution and disintegration of the working class took place, I revealed how everyday experiences mixed with the prevailing political and scientific powers, and how this mixing could reshape the working class, or prevent its renewal.

The first chapter, devoted to the politics of food, showed how much working-class reality was interconnected with the scientific knowledge of the times and the Habsburg Monarchy on the one hand, and how this knowledge and power influenced the shape of the working class on the other. The escalating supply crisis completely blurred Austrian society's prewar social structure. Industrial, qualified work, which in the past ensured at least a tolerable living wage, quickly transformed itself into a synonym of poverty, along with the whole urban environment within which this work was generally carried out. The critical deficiency of basic consumer goods led the Austrian state to implement a strict ration system that mostly affected urban salaried workers. When the working class's decreasing real income ceased to cover even the most basic necessities of life, many workers found themselves completely dependent on state, municipal, or company supply programs. They thus quickly came into everyday contact not only with official state power, but also with scientific knowledge, which claimed absolute authority in establishing alimentary requirements for the various social classes and in finding ways to satisfy those requirements.

The dominant scientific paradigm of the time, which equated workers' bodies with industrial motors, had a significant influence on probably the most important aspect of everyday wartime existence, i.e., on what, when, and how workers could eat. In fact, the wartime alimentary system did not respect prewar social stratification and to a certain extent created a completely new social hierarchy. Based on examples from the capital city of Prague, the largest Austrian munitions plant in Pilsen, and the northern Bohemian brown coal basin, we could see how these hierarchies bypassed the working class as a collective subject that united the providers of industrial labor. The critical food shortage broke the prewar working class in the realm of consumption, and scattered its members into many newly created groups. Qualified workers thus met the urban poor at the lowest rungs of the social hierarchy of the wartime cities. In middle-class public kitchens, however, they could become part of a collective that also consisted of traditionally notable urban members. Another segment of the working class relied on employers for food, which made it even more materially dependent than before the war. The least fortunate segment of the working class was left to rely on itself or stand in endless food lines.

Since working-class culture itself was significantly co-created by middle-class modern natural science discourse, the opportunities to extricate itself from the new organization of wartime consumption were greatly limited. Although food shortages gradually became a source of working-class dissatisfaction, the instances when the working-class environment generated its own alternative concepts to challenge the mainstream view of their nutritional requirements were rare. The working-class environment thus limited itself to protests against the fact that workers did not have enough food. The paradigm of modern nutritional science, which saw the providers of industrial labor merely in categories of the physical transfer of energy, was hardly questioned. The complex fantasy of the working body as a motor enabled the supply system to stay in place for a long time, during which increasing food shortages could be quickly satisfied and peace restored.

The examples of the Prague municipal kitchens or of the northern Bohemian brown coal basin reveal how little was needed for the prevention of more broadly articulated collective requirements. Even at the height of the crisis, the scattering of workers into several consumer groups, as in the Prague example, or into many canteens, as in the case of several dozen enterprises and mines in northern Bohemia, facilitated the rapid pacification of possible collective actions. The dominant metaphor of the working body as a motor not only informed the negotiations of the Austrian central bureaucracy or individual employers, but also of many workers. When human motors began to deteriorate due to a shortage of fuel, all it took was to procure the necessary nutrition as quickly as possible and the motors started running again. A total collapse of loyalty in the Austrian wartime industry occurred only in the last year of the war, when the central supply system was not able to satisfy even the most basic food requirements.

The politics of labor broke the working class as much as the politics of food. Here, too, the modern scientific discourse substituting working bodies with combustion engines played the main role. As in the case of food, in the sphere of industrial labor modern knowledge authoritatively determined how, when, and under what conditions one could work. It was the basic foundation of extensive transformations in the world of industrial labor, which completely changed the horizons of a vast majority of workers. The massive rationalization of the Austrian wartime industry deprived most workers of their personal autonomy. From their position as a contractual party on the labor market, they became the objects of rationalizational practices, applied by a new class of professional managers and engineers. In accordance with the ruling academic discourse, many workers were relegated to the same level as factory equipment.

Like combustion engines, workers' bodies also had clearly prescribed work—the time when they were supposed to be engaged in work and the time for rest and maintenance that allowed full capacity to be reached immediately afterwards. Just like in the sphere of food, here, too, the majority of workers were forced to submit to a rationing system. However, the commodity in question was not life-sustaining food, but the time that they had to spend at work or that they could devote to rest. Workers were also submitted to economizing logic, their bodies only seen as a variable in the equation of wartime production, similar to the output of combustion engines, heavy furnaces or automatic cranes. All of these factors together constituted the output of the Austrian wartime economy, and it was therefore necessary to have them working at full capacity. This was accompanied by restrictive legislation, based on the most modern scientific knowledge, and its enforcement, ushering previously unseen violence into the production halls of the Austrian industry.

Unlike the wartime politics of food, the wartime politics of labor not only shattered the organized working class but also subsequently led to its renewal. This went far beyond the mere maintenance of working-class culture's institutionalization in several large unions. Extensive changes in the sphere of industrial labor altered the horizon and working-class culture turned away from the prewar order as something that should be recovered during the postwar reconstruction. The traditional criticism that the organized socialist movement was heading toward a liberal arrangement of labor before the war gained credibility and became more and more acceptable to previously uninterested workers. By 1917, we can observe a significant rearrangement of perspectives in connection with the transformation of industrial labor when the working class began to orient itself towards the future, which it increasingly understood as radically different from the past. The working class's main enemy, however, was not the state, but employers. The utopian communist project, transforming from a parlor room topic of leftist theorists to a real social choice with significant working class support, gained relevance.

Although in both concluding chapters the motif of a "ration" is approached metaphorically, here, too, we can observe a decrease in abstract positive male self-identification on the one hand, and an increase in general dissatisfaction on the other. Although the cause was not the clearly tangible rations of food in chapter one, or the scientifically administered rations of work and rest in chapter two, we can also speak here about declining male dignity and, as the flipside of this, about increasing levels of anger that subsequently sent tens of thousands of workers into the streets of Austrian towns.

The collapse of positive male self-identification in the private space of the family, in the public space of the streets of Austrian towns, and in the production spaces of the Austrian industry further robbed the working class of its prewar position as an organized social agent. Its androcentric essence, which saw the male worker as the only subject that represented workers' interests, was shaken to its foundations. New scientific knowledge, which had a far-reaching impact on workers' everyday experiences, was in many cases accessible to working women and significantly undermined the former authority of the male worker as the main source of familial authority. The end of the prewar gender order in Austrian public space disrupted male workers' self-identification as respected representatives of working-class interests in the public sphere. The breakdown of the organic connection between public authority and the male gender shook up male workers' self-identification as bearers of rationality and public respect. The total deconstruction of the man as the prominent provider of hard industrial labor and social privileges associated with this status then completed the extinction of the working class as an androcentric collective with a clear male authority. The massive influx of female workers even into the most masculine industrial branches further strengthened the crisis of masculinity that male workers experienced as a consequence of wartime changes.

Similarly to the sphere of labor, the working class reconstituted itself in the sphere of gender as a new collective. The acceptance of women as full union members and their increasing infiltration into the prominent fora of Social Democracy as the most vocal defenders of working-class interests radically changed the working class from an androcentric collective into a mixed gender collective. Although in many cases it could be seen as the beginning of a process that was concluded much later, the working class in this sense experienced probably a faster and deeper transformation during the four war years than at any time during its history. The gradual involvement of women not only changed the basic quantitative proportions of both sexes, but also brought up new specific demands that further broadened the range of collectively raised claims. By opening itself up to women, the organized working class thus not only preserved its material base but also widened its demands to spheres that it had ignored before the war. Here, the radical demand for total political equal rights for all women, which in many successor states of the Habsburg Monarchy were granted immediately, is only one, albeit the most significant, example of how the transformation of the androcentric working-class collective into a mixed gender collective was reflected in its demands as well as in the composition of its supporters.

The influence of this expansion is also apparent in the last of the spheres of politics in question—the politics of protest. The final chapter focuses on two of its most visible forms, i.e. mass street unrests and organized factory strikes. Here, too, similarly to the sphere of labor, we see a fundamental turning point in the year 1917. While in several past works a strict dividing line was drawn between the two forms of wartime collective protest, if we look more closely at the protesting subjects, this line appears much more permeable; indeed, the riots were often the outcomes of industrial strikes.

One of the basic characteristics of such riots, regardless of how they came about, was an emphasis on an alternative understanding of justice derived from a specific "moral economy" that lay outside of market relations. This moral economy came out of the specific environment of the wartime industrial working class, where newly arrived workers, who came from the countryside, mingled with workers with more extensive industrial experiences. Capitalist formulas of protest, such as the organized refusal to work, mixed with non-capitalist forms, in which promoting the alternative "moral economy" through the authoritative enforcement of "fair prices" traditionally played one of the central roles.

Popular unrest, however, often did not just articulate disapproval of the middle-class capitalist understanding of justice. It also negated the other basic defining components of liberal urban society. Disrespect of private ownership, demonstratively extravagant treatment of looted plunder, and the protesting crowd's other seemingly senseless practices refuted the dominant liberal values of rationality and private ownership. The spatial aspects of the street riots often tried to at least temporarily occupy and symbolically recode the city centers that most symbolized the hegemony of political liberalism. Since the city in the modern societies of the nineteenth century developed into a single center of power, dominance over urban spaces meant dominance over the society as such. City centers and their "better addresses," where the private homes of the leaders of the ruling order were situated with their opulent liberal architecture representing the dominance of liberal elites over the given town, were the frequent targets of the looting crowd. Gaining control over the city center, albeit for a short while, or the short-term destabilization of the middle-class peace, thus meant questioning the central components of the entire society's inner stability.

Workers generally participated in street unrests—a wild type of popular protest—but they were rarely made up exclusively of workers. The primary type of working-class protest during World War I remained the organized strike. Although it could freely turn into a wilder street disturbance, in many cases this did not happen and many strikes were impressive examples of organized working-class protests. At the same time, several wartime strikes clearly revealed the limits of the newly constituted wartime working class, above all in cases when the strikers were not able to attract a larger segment of the industrial workers to their cause. Based on the examples of Pilsen and Prague, we have seen the most important ingredients that distinguished a successful strike from an unsuccessful one. While the Pilsen mass strike at the end of July and the beginning of August 1917 managed to integrate practically the entire staff of the Pilsen Škoda plant regardless of age, gender or nationality, the Prague strikes that took place at the same time never reached such proportions. The difference was in the very course of the strike protests, in their organization, and in the performative practices that maintained their cohesiveness. The mass concentration of human bodies was a significant and uncommon experience that informed the strike experience for many Pilsen workers, yet was entirely missing in Prague. A feeling of ideological, material, and also immediate physical togetherness thus ensured the success of the Pilsen strike, while the absence of such a feeling in the Prague case, where young, qualified, Czech-speaking men usurped the main role in the protest, was one of the main reasons why the Prague strikes were unsuccessful.

If, at the beginning of the book, the question was asked what influence the wartime context had on working-class culture, on the ways in which workers understood themselves and the world around them, or on the transformation of the organized working class, we must conclude that it was truly significant. Working-class culture, workers' self-positioning in the surrounding world and the very shape of the working class at the beginning of 1919 was entirely different from the situation in the fall of 1914. The seemingly depoliticized environment of wartime Austria created a collective agent that would become a stronger social force in the future with an incomparably greater influence on the surrounding world than before World War I. Thanks to its new openness to women, the organized working-class collective was more numerous than ever before. It had extensive knowledge of how far the reorganization of the world of work, which was slanted towards employers, could go. This knowledge enabled the formulation of radical demands that found fertile ground especially among radical left-wing workers and resulted in the foundation of the influential communist daily *Rudé Právo* in 1920 and eventually the independent Czechoslovak Communist party in 1921, which was about to play a significant role in the political life of the interwar Czechoslovakia. Although it faced intensifying waves of repression from the mid 1920s onwards, it still remained among the top vote-getters for seats in parliament throughout the interwar period. The World War I experience of constant repression might have played a role in the persisting influence of the persecuted party.

The postwar working class had empirically verified experience with the collective achievement of its demands. Mass strikes, which tried to integrate all of the working-class subjects regardless of age or gender, would thus become more frequent in the future. However strong the social influence of the male workers and their enfranchised female counterparts was in the interwar period, the reshaped working class never attained the prewar political unity. After the reestablishment of formal politics in the late Habsburg Monarchy in 1917, the chasm between the political leaders of the Social Democratic party and the rank and file became more and more apparent. Partly losing its influence in the fight with the Czech National Socialist party and partly not being able to unite moderate and radical Marxists, the claim to represent all working people could not be fulfilled by Social Democrats after 1918. Even though Social Democracy made a great effort to mobilize the workers for the cause of national independence in 1918, in the following years it could not claim to be the sole representative of workers' interests any more. Thus, actually, the Habsburg military dictatorship's attempt to dissipate the workers' political representation was, on its own, paradoxically successful.

The end of the war was not the end of the ration-based economy. The food ration system lasted until the beginning of the 1920s in many successor states, as did the food riots and other waves of violence. However, the idea of total organization lasted longer in Czech as well as in European thought. The fantasy of the absolute rationalization of the society, the production of their estates and their distribution returned like a boomerang during the twentieth century. The idea of a fully rationalized life, i.e., a "rationed life," where everything has a scientifically explained amount and purpose, and where there is therefore no room for any waste, did not disappear with World War I. In many aspects it survives to this day. Only the definition of what waste is, and how to prevent it, changes.

Bibliography

Archival Sources

Archives of the Capital City of Prague: Fond Ústřední sociální úřad (POM: 152).
Fond Okresní správa politická, Karlín (JAF č. 69).
Paměti Vojtěcha Bergera.
Referát XIV. Zásobování.
Central Military Archives in Prague: Fond 9. Sborové velitelství (1883–1919).
Labor Union Archives, Prague: Fond staré odborové spolky.

Published Archival Sources

Hájková, Dagmar, and Eva Kalivodová, eds. *Válečné deníky Edvarda a Hany Benešových (1915–1918)*. Prague, 2012.

Jemelka, Martin, ed. *Lidé z kolonií vyprávějí své dějiny*. Ostrava, 2009.

Neck, Rudolf, ed. *Arbeiterschaft und Staat im Ersten Weltkrieg I. Der Staat* (I. Vom Kriegsbeginn bis zum Prozess Friedrich Adler, August 1914–Mai 1917). Vienna, 1964.

———. *Arbeiterschaft und Staat im Ersten Weltkrieg I. Der Staat* (Vom Juni 1917 bis zum Ende der Donaumonarchie im November 1918). Vienna, 1968.

Otáhalová, Libuše, ed. *Souhrnná hlášení presidia pražského místodržitelství o protistátní, protirakouské a protiválečné činnosti v Čechách 1915–1918*. Prague, 1957.

Špiritová, Alexandra, ed. *Sborník dokumentů k vnitřnímu vývoji českých zemí za 1. světové války*, vol. IV. Prague, 1996.

Stupková, Marie, ed. *Sborník dokumentů k vnitřnímu vývoji českých zemí za 1. světové války*, vol. II. Prague, 1994.

Vrbata, Jaroslav, and Eva Drašarová, ed. *Sborník dokumentů k vnitřnímu vývoji českých zemí za 1. světové války*, vol. V. Prague, 1997.

Printed Primary Sources

Abderhalden, Emil. *Die Grundlagen unserer Ernährung unter besonderer Berücksichtigung der Jetztzeit*. Berlin, 1917.

Amar, Jules. *The Human Motor, or The scient. foundations of labour and industry*. London, 1927.

Bericht der k.k. Gewerbeinspektoren über Ihre Amtstätigkeit im Jahre 1914. Vienna, 1915.

Bericht der k.k. Gewerbeinspektoren über Ihre Amtstätigkeit im Jahre 1915. Vienna, 1916.

Bericht der k.k. Gewerbe-Inspektoren über ihre Amtstätigkeit im Jahre 1916. Vienna, 1919.

Dělnická kuchařka. Prague, 1914.

Fleischner, Jindřich. *Technická kultura. Sociálně-filosofické a kulturně-politické úvahy o dějinách technické práce*. Prague, 1916.

Gautier, Armand. *Diet and Dietetics*. London, 1906.

Habrman, Gustav. *Mé vzpomínky z války. Črty a obrázky o událostech a zápasech za svobodu a samostatnost*. Prague, 1928.

Holek, Václav. *Paměti: společná cesta české a německé sociální demokracie koncem devatenáctého století*. Prague, 2011.

Jahrbuch der österreichischen Industrie für das Jahr 1914.

Kafka, Josef. *Úsporná výživa. Hospodářská a zdravotní reforma výživy, jídelního lístku, vaření a zažívání*. Prague, 1915.

Kejřová, Anuše. *Dělnická kuchařka se zřetelem na malé dělnické domácnosti*. Hradec Králové, 1914.

———. *Kniha vzorné domácnosti: Vyzkoušené rady, pokyny a předpisy pro hospodyňky, jimž vzorné a úsporné vedení domácnosti na srdci leží*. Prague, 1916.

———. *Úsporná kuchařka: zlatá kniha malé domácnosti*. Hradec Králové, 1905.

Koropatnicki, Demeter. *Kommentar zum Kriegsleistungsgesetz: [vom 26.12. 1912] samt Nebengesetzen; in Verbindung mit dem Gesetzestext, den Erläuterungen des k.u.k. Kriegsministeriums, des Landesverteidigungsministeriums, den Beratungsprotokollen des Reichsrates etc*. Vienna, 1916.

Kraepelin, Emil. Die Arbeitskurve. Leipzig, 1902.

———. *O duševní práci*. Prague, 1906.

Lagrange, Ferdinand. *La médicacion par l´ escercice*. Paris, 1894.

Mareš, František. *Všeobecná fysiologie*. Prague, 1894.

———. *Výživa člověka ve světle fysiologie*. Prague, 1916.

———. *Fysiologie*, vol. I. Všeobecná fysiologie. Prague, 1906.

Marschner, Robert. *Die Fürsorge der Frauen für die heimkehrenden Krieger*. Prague, 1916.

Merhaut, Antonín. *Zásobování a výživa lidu v době válečné*. Prague, 1916.

Mosso, Angelo. *La Fatigua*. Milano, 1891.

———. *Tělesná výchova mládeže*. Prague, 1901.

Nedvěd, Antonín. *Jak to bylo na českém západě 1914–1918. Záznamy a dokumenty*. Pilsen, 1939.

Neruda, Jan. *Fejetony*. Prague, 2011.

Österreichischer Metallarbeiterverband. Bericht über die Tätigkeit des Verbandes in den Verwaltungsjahren 1914–1920. Vienna, 1921.

Pazourek, Josef. *Taylorova soustava organisace práce.* Královské Vinohrady, 1913.
Potocký, Pavel. *Dělník na válčených úkonech. Jeho práva a povinnosti.* Prague, 1915.
Reichsgesetzblatt für die im Reichsrate vertretenen Königreiche und Länder 1888, 1912, 1914–1917.
Richet, Charles. *L'Homme et l' Intelligence. Fragments de Physiologie et de Psychologie.* Paris, 1887.
Rubner, Max. *Die Gesetze des Energieverbrauchs bei der Ernährung.* Leipzig, 1902.
Špaček, Stanislav. *Práce a hospodářství: úvahy o lidské a pracovní ekonomii.* Prague, 1918.
Stenographise Protokolle des österreichischen Abgeordnetenhauses 1887.
Stoklasa, Julius. *Das Brot der Zukunft.* Jena, 1917.
———. *Výživa obyvatelstva ve válce!* Prague, 1916.
Taylor, Frederick W. *The Principles of Scientific Management.* New York and London, 1911.
Unie horníků rakouských, zpráva o činnosti za rok 1912, 1913, 1914, 1915, 1916 a 1917. Trnovany and Teplice, 1918.
von Voit, Carl. *Physiologie des allgemeinen Stoffwechsels und der Ernährung.* Leipzig, 1881.
Vznik a činnost kuchyní komitétu pro společné stravování méně majetného obyvatelstva v Praze v létech 1916–1920. Prague, 1920.
Weichardt, Wilhelm. *Über Ermüdungsstoffe.* Stuttgart, 1910.
Weyr, František. *Paměti. I. Za Rakouska (1879–1918).* Brno, 1999.
Yoteyko, Josefa. *Intróduction a la methodologie de la psychologie pédagogique.* Geneva, 1909.
Zuntz, Nathan. *Lehrbuch der Physiologie des Menschen.* Leipzig, 1909.

Periodical Press

Kovodělník. Orgán svazu dělnictva zaměstnaného výrobou a zpracováním kovů a drahokovů v Rakousku, 1914–1918.
Nová Doba 1914–1917.
Prager Tagblatt 1916.
Právo Lidu 1914–1917.
Zájmy žen. Časopis pro zájmy žen výdělečně pracujících, 1917.

Literature

Acton, Carol. *Grief in Wartime: Private Pain, Public Discourse.* London, 2007.
Altenhöner, Florian. *Kommunikation und Kontrolle: Gerüchte und städtische Öffentlichkeiten in Berlin und London 1914/1918.* Munich, 2008.
Augeneder, Sigrid. *Arbeiterinnen im Ersten Weltkrieg. Leben- und Arbeitsbedingungen proletarischer Frauen in Österreich.* Vienna, 1987.

Bahenská, Marie. "Pomalu, pozvolna, po špičkách. K chápání a reflexi pojmu emancipace v českých zemích v 19. a 20. století." *Moderní dějiny*, suppl. 1 (2008): 444–57.

Bahenská, Marie, Libuše Heczková and Dana Musilová, ed. *Ženy na stráž! České feministické myšlení 19. a 20. století*. Prague, 2010.

Bauerkämper, Arnd and Elise Julien, ed. *Durchhalten! Krieg und Gesellschaft im Vergleich 1914–1918*. Göttingen, 2010.

Belzer-Scardino, Allison. *Women and the Great War: Femininity Under Fire in Italy*. New York, 2010.

Beneš, Jakub. "Czech Social Democracy, František Soukup, and the Habsburg Austrian Suffrage Campaign 1897–1907: Toward a New Understanding of Nationalism in the Workers' Movements of East Central Europe." *Střed/Centre* 2 (2012): 9–33.

———. "Socialist Popular Literature and the Czech-German Split in Austrian Social Democracy 1890–1914." *Slavic Review* 72, no. 2 (2013): 327–51.

Benevolo, Leonardo. *Die Stadt in der Europäischen Geschichte*. Munich, 1993.

Benko, Juraj. *Boľševizmus medzi východom a západom (1900–1920)*. Bratislava, 2012.

Berger, Stefan. "Die europäische Arbeiterbewegung und ihre Historiker: Wandlungen und Ausblicke." *Jahrbuch für europäische Geschichte* 6 (2005): 151–82.

Berlin, Isaiah. *Čtyři eseje o svobodě*. Prague, 1999.

Bevilacqua, Fabio. "Helmholtz's Über die Erhaltung der Kraft: The Emergence of a Theoretical Physicist." In *Hermann von Helmholtz and the Foundations of Nineteenth-Century Science*, edited by David Cahan, Berkeley, 1994, 291–333.

Bluma, Lars, and Karsten Uhl, eds. *Kontrollierte Arbeit—Disziplinierte Körper? Zur Sozial- und Kulturgeschichte der Industriearbeit im 19. und 20. Jahrhundert*. Bielefeld, 2012.

Böhnisch, Lothar. *Männliche Sozialisation: Eine Einführung*. Weinheim, 2004.

Bourke, Joanna. *Dismembering the Male: Men's Bodies, Britain and the Great War*. Chicago, 1996.

Brass, Tom, and Marcel van der Linden, ed. *Free and Unfree Labour: The Debate Continues*. Bern and Frankfurt am Main, 1997.

Bruch, Rüdiger vom. "Streiks und Konfliktregelung im Urteil bürgerlicher Sozialreformer 1872–1914." In *Streik. Zur Geschichte des Arbeitskampfes in Deutschland während der Industrialisierung*, edited by Klaus Tenfelde and Heinrich Volkmann, Munich, 1981, 253–70.

Brückweh, Kerstin, Dirk Schumann, Richard F. Wetzell, and Benjamin Ziemann, eds. *Engineering Society: The Role of the Human and Social Sciences in Modern Societies, 1880–1980*. Basingstoke, 2012.

Burawoy, Michael. "The Anthropology of Work." *Annual Review of Anthropology* 8 (1979): 231–66.

Bürgschwentner, Joachim, Mattthias Egger, and Gunda Barth-Scalmani, eds. *Other Fronts, Other Wars? First World War Studies on the Eve of the Centennial*. Leiden and Boston, 2014.

Canetti, Elias. *Masa a moc*. Prague, 1994.

Canning, Cathleen. *Gender History in Practice. Historical Perspectives on Bodies, Class, and Citizenship*. Ithaca and London, 2006.

Cardwell, Donald S. L. *James Joule: A Biography*. Manchester and New York, 1989.

Cattaruzza, Marina. *Sozialisten an der Adria. Plurinationale Arbeiterbewegung in der Habsburgermonarchie*. Berlin, 2011.

Čepelák, Václav, ed. *Dějiny Plzně II. Od roku 1788 do roku 1918*. Pilsen, 1967.

Chakrabarty, Dipesh. *Provincilizing Europe: Postcolonial Thought and Historical Difference*. Princeton, NJ, 2000.

Chickering, Roger, and Stig. Förster, eds. *Great War, Total War, Combat and Mobilization on the Western Front, 1914–1918*. Cambridge, 2006.

Clark, Christopher. *The Sleepwalkers: How Europe Went to War in 1914*. New York, 2012.

Cohen, Deborah. *The War Came Home: Disabled Veterans in Britain and Germany 1914–1939*. Berkeley, CA, 2001.

Connell, Raewyn W. "The Big Picture: Masculinities in Recent World History." *Theory and Society* 22, no. 5 (1993): 507–44.

———. *Masculinities*. Cambridge, 2005.

Cott, James C. *The Moral Economy of the Peasant: Rebellion and Subsistence in Southeastern Asia*. New Haven, CT and London, 1976.

Cuřínová, Ludmila. "Ústav pro národní eugeniku." In *Technokracie v českých zemích (1900–1950)*, edited by Jan Janko and Emilie Těšínská, Prague, 1999, 151–156.

Custer, Paul A. "Refiguring Jemima: Gender, Work and Politics in Lancashire 1770–1820." *Past & Present* 195 (2007): 126–58.

Czarnowski, Gabriele, and Elisabeth Meyer-Renschhausen. "Geschlechterdualismen in der Wohlfahrtspflege: Soziale Mütterlichkeit zwischen Professionalisierung und Medikalisierung." *L'Homme. Zeitschrift für feministische Geschichtswissenschaft* 2, no. 5 (1994): 121–40.

Das, Santanu. "Kiss Me, Hardy: Intimacy, Gender, and Gesture in First World War Trench Literature." *Modernism/Modernity* 9, no. 1 (2002): 51–74.

Davidoff, Leonore, and Catherine Hall. *Family Fortunes. Men and Women of the English Middle Class 1780–1850*. London, 2002.

Davis, Belinda J. *Home Fires Burning. Food, Politics and Everyday Life in World War I Berlin*. Chapel Hill, NC and London, 2000.

Daunton, Martin, and Matthew Dilton, eds. *The Politics of Consumption: Material Culture in Europe and America*. New York and Oxford, 2001.

De Grazia, Victoria, ed. *The Sex of Things: Gender and Consumption in Historical Perspective*. Berkeley, 1996.

Deak, John "The Grat War and the Forgotten Realm: The Habsburg Monarchy and the First World War." *Journal of Modern History* 86, no. 2 (June 2014): 336–80.

Diederichs, Helmut H. *Frühgeschichte deutscher Filmtheorie. Ihre Entstehung und Entwicklung bis zum Ersten Weltkrieg*. Frankfurt am Main, 1996.

Domansky, Elizabeth. "Militarization and Reproduction in World War I Germany." In *Society, Culture, and the State in Germany, 1870–1930*, edited by Geoff Eley, Ann Arbor, MI, 1996, 427–464.

———. "The rationalization of class struggle: strikes and strike strategy of the German Metalworkers' Union, 1892–1922," In *Strikes, Wars, and Revolutions in International Perspective: Strike Waves in the Late Nineteenth and Early Twentieth Century*, edited by Leopold D. Haimson and Charles Tilly, Cambridge and New York, 2002, 321–355.

Drobesch, Werner. "Die soziale Frage der Habsburgermonarchie im zeitgenössischen gesellschaftswissenschaftlicen Diskurs." *Moderní dějiny/Modern History* 1 (2012): 1–12.

———. "Ideologische Konzepte zur Lösung der sozialen Frage." In *Die Habsburgermonarchie 1848–1918*, vol. IX, edited by Ulrike Harmat, Vienna, 2010, 1419–1463.

Druhý list Pavla Tesalonickým 3, 10. Bible. Prague, 2012.

Ebert, Kurt. *Die Anfänge der der modernen Sozialpolitik in Österreich. Die Taafesche Sozialgesetzgebung für die Arbeiter im Rahmen der Gewerbeordnungsreform (1879–1885).* Vienna, 1975.

Efmertová, Marcela C. *Elektrotechnika v českých zemích a v Československu do poloviny 20. století.* Prague, 1999.

Ehlert, Hans Gotthard. *Die wirtschaftlichen Zentral-behörden des Deutschen Reiches 1914 bis 1919. Das Problem der "Gemeinwirtschaft" in Krieg und Frieden.* Wiesbaden, 1982.

Eisenhammer, Miroslav. "Mzdová hladina zaměstnanců Škodových závodů v Plzni v době první světové války." *Historie a vojenství* 4 (2000): 766–79.

Eley, Geoff, and Keith Nield. "Farewell to the Working Class?" *International Labor and Working Class History* 5 (2000): 1–30.

———. *The Future of Class in History. What´s Left of the Social?* Ann Arbor, 2007.

Englová, Jana. "Dělnictvo jako subjekt a objekt historického bádání." In *Problematika dělnictva v 19. a 20. století. Bilance a výhledy studia*, edited by Stanislav Knob and Tomáš Rucki Ostrava, 2011, 34–39.

Etzemüller, Thomas. "Social Engineering als Verhaltenslehre des kühlen Kopfes. Eine einleitende Skizze." In *Die Ordnung der Moderne. Social Engineering im 20. Jahrhundert*, edited by Thomas Etzemüller, Bielefeld, 2009, 11–40.

Fasora, Lukáš. *Dělník a měšťan. Vývoj jejich vzájemných vztahů na příkladu šesti moravských měst 1870–1914.* Brno, 2010.

———. "Generační revolta v socialistickém táboře v letech 1900–1920." *Český časopis historický* 2 (2012): 288–317.

Fasora, Lukáš, Jiří Hanuš, and Jiří Malíř, ed. *Sozial-reformatorisches Denken in den böhmischen Ländern 1848–1914.* Munich, 2010.

Folbre, Nancy. *Greed, Lust and Gender: A History of Economic Ideas.* Oxford and New York, 2009.

Foucault, Michel. *Power/Knowledge: Selected Interviews and Other Writings, 1972–1977.* New York, 1980.

Frader, Laura. "Dissent Over Discourse. Labor History, Gender and the Linguistic Turn." *History and Theory* 34 (1995): 213–30.

Franc, Martin. "Výmysly německých profesorů i návrat k zkušenostem předků. Přírodní potravinové náhražky za 1. světové války v českých zemích." *Práce z dějin Akademie věd* 4, no. 1 (2012): 1–14.

Frévert, Ute, ed. *Bürgerinnen und Bürger. Geschlechterverhältnisse im 19. Jahrhundert.* Göttingen, 1988.

Gagnier, Regenia. *Subjectivities: A History of Self-Representation in Britain, 1832–1920.* New York and Oxford, 1991.

Galandauer, Jan. *Bohumír Šmeral 1914–1941.* Prague, 1986.

———. *Ohlas Velké říjnové socialistické revoluce v české společnosti.* Prague, 1977.

Gelnarová, Jitka. "Matka Praha a dcery její. Diskuse o ženském volebním právu do obce pražské v občanském a dělnickém ženském hnutí mezi lety 1906 a 1909." *Střed/Centre* 2 (2011): 34–58.

Geršlová, Jana. "Elektroindustrie in den böhmischen Ländern: Emil Kolben und die Entwicklung der Firma ČKD Prag." In *Business History. Wissenschaftliche Entwicklungstrends und Studien aus Zentraleuropa,* edited by Alice Teichová, Herbert Matis, and Andreas Resch, Vienna, 1999, 197–203.

Geyer, Michael. "The Militarization of Europe, 1914–1945." In *The Militarization of the Western World,* edited by J. Gillis, New Brunswick, NJ, 1989, 74–79.

Goebel, Stefan, and Derek Keene, eds. *Cities into Battlefields: Metropolitan Scenarios, Experiences and Commemorations of Total War.* Farnham, 2011.

Goldstein, Joshua S. *War and Gender: How Gender Shapes the War System and Vice Versa.* Cambridge and New York, 2004.

Grobelný, Andělín, and Bohumil Sobotík. *Dělnické hnutí na Ostravsku.* Ostrava, 1957.

Guerin, Frances. *Culture of Light: Cinema and Technology in 1920s Germany.* Minneapolis, 2005.

Haas, Hanns, and Hannes Stekl, eds. *Bürgerliche Selbstdarstellung: Städtebau, Architektur, Denkmäler.* Vienna, 1995.

Haimson, Leopold D., and Charles Tilly, eds. *Strikes, Wars, and Revolutions in International Perspective: Strike Waves in the Late Nineteenth and Early Twentieth Century.* Cambridge and New York, 2002.

Hämmerle, Christa. *Heimat/Front. Geschlechtergeschichte(n) des Ersten Weltkriegs in Österreich-Ungarn.* Vienna, 2014.

Hämmerle, Christa, Oswald Überegger, and Brigitta Bader Zaar, eds. *Gender and the First World War. Basingstoke,* 2014.

Hanusch, Ferdinand, and Emanuel Adler, eds. *Die Regelung der Arbeitsverhältnisse im Kriege.* Vienna and New Haven, 1927.

Hausen, Karin. "Family and Role Division: The Polarisation of Sexual Stereotypes in the Nineteenth Century—An Aspect of the Dissociation of Work and Family Life." In *The German Family: Essays on the Social History of the Family in Nineteenth and Twentieth-Century Germany,* edited by Richard J. Evans and Robert W. Lee, 51–83. London, 1981.

Hautmann, Hans. "Hunger ist ein schlechter Koch. Die Ernährungslage der österreichischen Arbiter im Ersten Weltkrieg." In *Bewegung und Klasse. Studien zur österreichischen Arbeitergeschichte,* edited by Gerhard Botz, Vienna, 1978, 661–682.

Havránek, Jan. "Politické represe a zásobovací potíže v českých zemích v letech 1914–1918." In *První světová válka a vztahy mezi Čechy, Slováky a Němci,* edited by Hans Mommsen, Dušan Kováč, and Jiří Malíř, Brno, 2000, 37–52.

Healy, Maureen. "Becoming Austrian: Women, the State, and Citizenship in World War I," *Central European History* 35, No. 1 (2002): 1–35.

———. "Civilizing the Soldier in Postwar Austria." In *Gender and War in Twentieth Century Eastern Europe,* edited by Nancy M. Wingfield and Maria Bucur, Bloomington, IN, 2006, 47–69.

———. *Vienna and the Fall of the Habsburg Empire: Total War and Everyday Life in World War I.* Cambridge and New York, 2004.

Heczková, Libuše, and Kateřina Svatoňová. "Úvod: Nebezpečná Božena Viková-Kunětická." In *Jus Suffragii. Politické projevy Boženy Vikové-Kunětické z let 1890–1926*, edited by Libuše Heczková and Kateřina Svatoňová, Prague, 2012, 7–19.

Herzog, Dagmar. *Sexuality in Europe: A Twentieth-Century History*. Cambridge and New York, 2011.

Heumos, Peter. "Kartoffeln her oder es gibt eine Revolution. Hungerkrawalle, Streiks und Massenproteste in den böhmischen Ländern 1914–1918." *Slezský sborník* 97, no. 2 (1999): 81–104.

Hierholzer, Viera. *Nahrung nach Norm. Regulierung von Nahrungsmittelqualität in der Industrialisierung 1871–1914*. Göttingen, 2010.

Hitzer, Bettina, and Thomas Welskopp, eds. *Die Bielefelder Sozialgeschichte. Klassische Texte zu einem geschichtswissenschaftlichen Programm*. Bielefeld, 2010.

Hobsbawm, Eric J. *Labouring Men: Studies in the History of Labour*. London, 1965.

Holubec, Stanislav. *Lidé periferie. Sociální postavení a každodennost pražského dělnictva v meziválečné době*. Pilsen, 2009.

Homburg, Heidrun. *Rationalisierung und Industriearbeit. Arbeitsmarkt—Management—Arbeiterschaft im Siemens-Konzern Berlin 1900–1939*. Berlin, 1991.

Horne, John, ed. *State, Society, and Mobilization in Europe during the First World War*. Cambridge, 1997.

Houfek, Václav. "Nacionalizace společnosti a dělnictvo na Ústecku do roku 1918." In *Nacionalizace společnosti v Čechách 1848–1914*, edited by Kristina Kaiserová and Jiří Rak, Ústí nad Labem, 2008, 281–315.

Hunt, Lynn. *Politics, Culture, and Class in the French Revolution*. Berkeley and London, 1984.

Hwaletz, Otto. "Österreichisch-Alpine Montangesellschaft bis in die 1930er Jahre." In *Business History. Wissenschaftliche Entwicklungstrends und Studien aus Zentraleuropa*, edited by Alice Teichová, Herbert Matis, and Andreas Resch, Vienna, 1999, 229–246.

Janáček, František. *Největší zbrojovka monarchie. Škodovka v dějinách, dějiny ve Škodovce 1859–1918*. Prague, 1990.

Janderová, Helena. "Psychotechnický ústav," in *Technokracie v českých zemích (1900–1950)*, edited by Jan Janko and Emilie Těšínská, Prague, 1999, 135–143.

Jemelka, Martin. *Na Šalomouně: společnost a každodenní život v největší moravskoostravské hornické kolonii (1870–1950)*. Ostrava, 2008.

Jindra, Zdeněk. *První světová válka*. Prague, 1984.

Jíša, Václav. *Škodovy závody 1859–1919*. Prague, 1965.

Joas, Hans. "Kontingenzbewusstsein: der Erste Weltkrieg und der Bruch im Zeitbewusstsein der Moderne." In *Aggression und Katharsis: der Erste Weltkrieg im Diskurs der Moderne*, edited by Petra Ernst, Sabine Haring, and Werner Suppanz, Vienna, 2004, 43–56.

Johannisson, Karin. "Modern Fatigue: A Historical Perspective." In *Stress in Health and Disease*, edited by Bengt B. Arnetz, and Rolf Ekman Weinheim, 2006), 3–19.

Jones, Geoffrey, ed. *Coalitions and Collaboration in International Business*. Adlershot, 1993.

Kann, Mark E. *The Gendering of American Politics: Founding Mothers, Founding Fathers, and Political Patriarchy*. London, 1999.

Kantorowicz, Ernst. *The King's Two Bodies: A Study in Mediaeval Political Theology*. Princeton, 1957.

Kárník, Zdeněk. *Habsburk, Masaryk či Šmeral. Socialisté na rozcestí*. Prague, 1996.

Kárný, Miroslav et al. *Sto let Kladenských železáren. Příspěvek k dějinám českého železářství a k dějinám dělnického hnutí na Kladensku v letech 1854–1957*. Prague, 1959.

Kaster, Gregory L. "Labor's True Man: Organized Workingmen and the Language of Manliness in the USA, 1827–1877." *Gender and History* 1 (2001): 24–64.

Kaufmann, Bruce. *Managing the Human Factor: The Early Years of Human Resource Management in American Industry*. Ithaca and London, 2008.

Kennan, George F. *The Decline of Bismarck's European Order: Franco-Russian Relations 1875–1890*. Princeton, NJ, 1979.

Kerber, Linda. "Separate Spheres, Female Worlds, Woman's Place: The Rhetoric of Women's History." *Journal of American History* 75, no. 1 (1988): 9–39.

Kessler-Harris, Alice. "Treating the Male as Other. Re-defining the Parameters of Labor History." *Labor History* 34 (1993): 190–204.

Kienitz, Sabine. "Body Damage: War Disability and Constructions of Masculinity in Weimar Germany." In *Home/Front: The Military, War and Gender in 20th Century Germany*, edited by Karen Hagemann and Stefania Schüler Springorum, New York and Oxford, 2002, 181–204.

Klages, Helmut. "Planung—Entwicklung—Entscheidung: Wird die Geschichte herstellbar?" *Historische Zeitschrift* 226 (1978): 529–46.

Kleinschmidt, Christian. *Rationalisierung als Unternehmersstrategie. Die Eisen- und Stahlindustrie des Ruhrgebietes zwischen Jahrhundertwende und Weltwirtschaftskrise*. Essen, 1993.

Klíma, Karel et al. *100 let ocelí Poldi*. Kladno, 1989.

Knob, Stanislav. "Stávkové hnutí v československé i zahraniční historiografii, srovnání a výhledy." In *Problematika dělnictva v 19. a 20. století. Bilance a výhledy studia*, edited by Stanislav Knob and Tomáš Rucki, Ostrava, 2011, 65–72.

Kocka, Jürgen. *Arbeitsverhältnisse und Arbeiterexistenzen, Grundlagen der Klassenbildung im 19. Jahrhundert*. Bonn, 1990.

———. *Klassengesellschaft im Krieg. Deutsche Sozialgeschichte 1914–1918*. Göttingen, 1973.

Kolář, Pavel, and Michal Kopeček. "A Difficult Quest for New Paradigms: Czech Historiography After 1989." In *Narratives Unbound. Historical Studies in Post-Communist Eastern Europe*, edited by Sórin Antohi, Balász Trencsényi, and Peter Apor, Budapest and New York, 2006, 173–248.

Kolejka, Josef, ed. *Z dějin dělnického hnutí na Brněnsku. Od nejstarších počátků do založení KSČ*. Brno, 1956.

———. *Revoluční dělnické hnutí na Moravě a ve Slezsku 1917–1921*. Prague, 1957.

Kolejka, Josef, and Václav Peša. "V boji proti válce i proti monarchii 1917–1918." In *Z dějin dělnického hnutí na Brněnsku. Od nejstarších počátků do založení KSČ*, edited by Josef Kolejka, Brno, 1956, 144–155.

Konrad, Helmut. "Arbeiterbewegung und bürgerliche Öffentlichkeit. Kultur und nationale Frage in der Habsburgermonarchie." *Geschichte und Gesellschaft* 20, no. 4 (1994): 506–18.

Kořalka, Jiří. *Češi v habsburské Říši a v Evropě 1815–1914. Sociálněhistorické souvislosti vytváření novodobého národa a národnostní otázky v českých zemích.* Prague, 1996.

Kramer, Alan. *The Dynamics of Destruction: Culture and Mass Killing in the First World War.* Oxford, 2008.

———. "Recent Historiography of the First World War. Part I." *The Journal of Modern European History* 12, no. 1 (2014): 5–27.

———. "Recent Historiography of the First World War. Part II." *The Journal of Modern European History* 12, no. 2 (2014): 155–74.

Kraus, Karl. *Die letzen Tage der Menschheit.*Vienna, 1918.

Křen, Jan. *Dvě století střední Evropy.* Prague, 2006.

Křepeláková, Vlastimila. *Struktura a sociální postavení dělnické třídy v Čechách 1906–1914.* Prague, 1974.

Krivanec, Eva. *Kriegsbühnen: Theater im Ersten Weltkrieg. Berlin, Lissabon, Paris und Vienna.* Bielefeld, 2012.

Křížek, Juraj. "Die Kriegswirtschaft und das Ende der Monarchie," In *Die Auflösung des Habsburgerreiches. Zusammenbruch und Neuorientierung im Donauraum,* edited by Richard Georg Plaschka, and Karlheinz Mack, Munich, 1970, 43–52.

Kučera, Rudolf. "Marginalizing Josefina: Work, Gender and Protest in Bohemia 1820–1844." *Journal of Social History* 46, no. 2 (2012): 430–48.

Kuleman, Peter. *Am Beispiel des Austromarxismus. Sozialdemokratische Arbeiterbewegung in Österreich von Hainfeld bis zur Dollfuß-Diktatur.* Hamburg, 1979.

Lacina, Vlastislav. "Válečné hospodářství ve střední Evropě a v českých zemích za první a druhé světové války." In *Česká společnost za velkých válek 20. století. Pokus o komparaci,* edited by Jan Gebhart and Ivan Šedivý, Prague, 2003, 45–50.

Landes, Joan, ed. *Feminism, the Public and the Private.* Oxford, 1998.

Langewiesche, Dieter. *Zur Freizeit des Arbeiters. Bildungsbestrebungen und Freizeitgestaltung österreichischer Arbeiter im Kaiserreich und der Ersten Republik.* Stuttgart, 1979.

Latour, Bruno. *We Have Never Been Modern.* Cambridge, MA, 1993.

Lawrence, Jon. "Forging a Peaceable Kingdom: War, Violence, and Fear of Brutalization in Post-First World War Britain." *Journal of Modern History* 75, no. 3 (2003): 557–89.

Leonhard, Jörn. *Die Büchse der Pandora. Geschichte des Ersten Weltrkieges.* München, 2014.

Lerner, Paul. *Hysterical Men: War Psychiatry and the Politics of Trauma in Germany 1890–1930.* Ithaca, 2003.

Lieffers, Caroline. "The Present Time is Eminently Scientific: The Science of Cookery in Nineteenth-Century Britain." *Journal of Social History* 45, no. 4 (2012): 936–59.

Lieske, Adina. *Arbeiterkultur und bürgerliche Kultur in Pilsen und Leipzig.* Bonn, 2007.

Linden, Marcel van der. *Workers of the World. Essays Towards a Global Labour History.* Leiden and Boston, 2008.

Loewenfeld-Russ, Hans. *Die Regelung der Volksernährung im Kriege.* Vienna and New Haven, 1926.

Loiperdinger, Martin, ed. *Filmpionier der Kaiserzeit*. Basel, 1994.

Löw, Raimund. "Die Deutsche Sozialdemokratie in Österreich und die Balkankriege 1912/1913." In *Internationalism in the Labour Movement 1830–1940*, edited by Frits von Holthoon and Marcel van der Linden, Leiden, 1988, 410–439.

Machačová, Jana. *Mzdové a stávkové hnutí dělnictva v průmyslových oblastech českých zemí 1848–1914*. Opava, 1990.

———. "Výzkum stávek z 19. a 20. století v československé historiografii. Přístupy českých a slovenských historiků." *Studie k sociálním dějinám* 2 (1998): 139–51.

Machačová, Jana, and Jiří Matějček. "Chudé (dolní) vrstvy společnosti českých zemí v 19. století. Sociální pozice a vzory chování." *Studie k sociálním dějinám* 1 (1998): 121–303.

———. *Nástin sociálního vývoje českých zemí 1781–1914*. Prague, 2010.

Maderthaner, Wolfgang, and Lutz Musner. *Unruly Masses: The Other Side of Fin-de-Siécle Vienna*. New York and London, 2008.

Maier, Charles S. "Between Taylorism and Technocracy: European Ideologies and the Vision of of Industrial Productivity in the 1920s." *Journal of Contemporary History* 5 (1970): 27–61.

Malínská, Jana. "Volební právo žen do říšské rady, českého zemského sněmu a obcí v letech 1848–1914." *Střed/Centre* 1 (2009): 24–57.

Marek, Michaela. "Monumentalbauten und Städtebau als Spiegel des gesellschaftlcihen Wandels in der 2. Hälfte des 19. Jahrhunderts." In *Böhmen im 19. Jahrhundert*, edited by Ferdinand Seibt, Frankfurt am Main, 1995, 149–223.

Mareš, Jan, and Vít Strobach. "Třída dělníků i žen? Proměny chápání genderových vztahů v českém dělnickém hnutí (1870–1914)." *Střed/Centre* 2 (2012): 34–68.

Martínek, Miloslav. "Přehled vývoje rakouského zákonodárství v oblasti chudinství, zdravotnictví a sociální správy." *Sborník k dějinám 19. a 20. století* 4 (1977): 63–85.

Matějček, Jiří. "Dělnické hnutí v Českých zemích do roku 1914. Emancipace dělnictva, nebo hegemonie proletariátu? Pokus o objektivní hodnocení vývoje hnutí i stavu výzkumu." *Studie k sociálním dějinám* 2 (1998): 153–95.

May, Arthur J. *The Pasing of the Hapsburg Monarchy 1914–1918*, vol. II. Philadelphia, 1966.

Melton, James Van Horn. *The Rise of the Public in Enlightenment Europe*. Cambridge and New York, 2001.

Mergel, Thomas, and Christiane Reinecke, eds. *Das Soziale ordnen. Sozialwissenschaften und gesellschaftliche Ungleichheit im 20. Jahrhundert*. Frankfurt am Main, 2012.

Mesch, Michael. *Arbeiterexistenz in der Spätgründerzeit—Gewerkschaften und Lohnentwicklung in Österreich 1890–1914*. Vienna, 1984.

Meyer, Jessica. *Man of War: Masculinity and the First World War in Britain*. Basingstoke, 2009.

Mommsen, Wolfgang J. Die Urkatastrophe Deutschlands. Der Erste Weltkrieg 1914–1918. Stuttgart, 2002.

Montanari, Massimo. *Hlad a hojnost. Dějiny stravování v Evropě*. Prague, 2003.

Mosse, George L. *Fallen Soldiers: Reshaping the Memory of the World Wars*. New York and Oxford, 1990.

Moutvic, Miroslav. *Ústřední jatky města Prahy v Holešovicích: 1895–1951*. Prague, 2007.

Müller, Ingo. *A History of Thermodynamics: The Doctrine of Energy and Entropy*. Berlin, 2007.

Münkler, Herfried. *Der Große Krieg. Die Welt 1914–1918*. Berlin, 2013.

Nelson, Daniel. *Managers and Workers: Origins of New Factory System in the United States 1880–1920*. Cambridge, MA, 1977.

Neve, Monica. *Sold! Advertising and the Bourgeois Female Consumer in Munich 1900–1914*. Stuttgart, 2010.

Niedermayr, Gerhard, and Franz Pertlik. "Hans J. Karabacek. Ein später Nachruf." *Mitteilungen der Österreichischen Mineralgesellschaft* 145 (2000): 15–20.

O'Quinn, Kimberly. "The Reason and Magic of Steel: Industrial and Urban Discourses in DIE POLDIHÜTTE." In *A Second Life: German Cinema's First Decades*, edited by Thomas Elsaesser, Amsterdam, 1996, 192–201.

Osietzki, Maria. "Körpermaschinen und Dampfmaschinen. Vom Wandel der Physiologie und des Körpers unter dem Einfluß von Industrialisierung und Thermodynamik." In *Physiologie und industrielle Gesellschaft. Studien zur Verwissenschaftlichung des Körpers im 19. und 20. Jahrhundert*, edited by Philipp Sarasin and Jakob Tanner, Frankfurt am Main, 1998, 313–346.

Otruba, Gustav. "Entstehung und soziale Entwicklung der Arbeiterschaft und der Angestellten bis zum Ersten Weltkrieg." In *Österreichs Sozialstrukturen in historischer Sicht*, edited by Erich Zöllner, Vienna, 1980, 123–154.

Pelikánová, Jaroslava et al. *Přehled dějin Československého odborového hnutí*. Prague, 1984.

Pepper, Hugo. "Die frühe österreichische Sozialdemokratie und die Anfänge der Arbeiterkultur." In *Sozialdemokratie und Habsburgerstaat*, edited by Wolfgang Maderthaner, Vienna, 1988, 79–100.

Perry, Heather. *Recycling the Disabled: Army, Medicine and Modernity in WWI Germany*. Manchester, 2014.

Plaschka, Richard Georg, Horst Hasselsteiner, and Arnold Suppan. *Innere Front. Militärassistenz, Widerstand und Umsturz in der Donaumonarchie 1918*, vols. I–II. München, 1974.

Pohl, Hans, ed. *Kartelle und Kartellgesetzgebung in Praxis und Rechtssprechung vom 19. Jahrhundert bis zur Gegenwart*. Stuttgart, 1985.

Polišenský, Josef, Karel Novotný, and Věra Vomáčková. *Boj dělníků na stavbách našich prvních železnic*. Prague, 1956.

Průcha, Vladimír. "Nástin vývoje nominální mzdy zaměstnaného průmyslového dělníka v Československu v letech 1913–1937." *Sborník historický* 13 (1965): 65–91.

Pullmann, Michal. "Revoluce a utváření nového: Vídeň, Praha a Berlín kolem r. 1918. K hodnotovým aspektům tří revolucí ve střední Evropě." Ph.D. diss., Charles University, 2002.

Purš, Jaroslav. *Postavení dělnické třídy a stávkové boje v českých zemích v období průmyslové revoluce*. Prague, 1964.

Rabinbach, Anson. "Ermüdung, Energie und der menschliche Motor." In *Physiologie und industrielle Gesellschaft. Studien zur Verwissenschaftlichung des Körpers im

19. und 20. Jahrhundert, edited by Philipp Sarasin and Jakob Tanner, Frankfurt am Main, 1998, 286–312.

———. "The European Science of Work: The Economy of the Body at the End of the Nineteenth Century." In *Work in France: Representations, Meanings, Organization, and Practice*, edited by Steven L. Kaplan and Cynthia J. Koepp, Ithaca and London, 1984, 475–513.

———. *The Human Motor: Energy, Fatigue and the Origins of Modernity*. Berkeley and Los Angeles, 1992.

Rahikainen, Marjatta. *Centuries of Child Labour: European Experiences from Seventeenth to the Twentieth Centuries*. Adlershot, 2004.

Rákosník, Jakub. *Odvrácená tvář meziválečné prosperity. Nezaměstnanost v Československu v letech 1918–1938*. Prague, 2008.

Řehák, Bohumil. "Za první války ve Škodovce." In *K dějinám závodů V. I. Lenina*, edited by Václav Jíša and Jiří Kodeš, Pilsen, 1962, 165–173.

Roberts, Mary M. *Civilization Without Sexes: Reconstructing Gender in Postwar France, 1917–1927*. Chicago, 1994.

Roper, Michael. *The Secret Battle: Emotional Survival in the Great War*. Manchester, 2010.

Sablik, Karl. *Julius Tandler. Mediziner und Sozialreformer*. Vienna, 1983.

Sachße, Christoph. *Mütterlichkeit als Beruf. Sozialarbeit, Sozialreform und Frauenbewegung 1871–1929*. Weinheim, 2003.

Sarasin, Philipp. *Reizbare Maschinen, Eine Geschichte des Körpers 1750–1914*. Frankfurt am Main, 2001.

Scheufler, Pavel. "Zásobování potravinami v Praze v letech 1. světové války." *Etnografie dělnictva* 9 (1977): 143–97.

Schlegel-Matthies, Kirsten. *Im Haus und am Herd. Der Wandel des Hausfrauenbildes und der Hausarbeit 1880–1930*. Stuttgart, 1995.

Schröer, Heinz, ed. *Carl Ludwig. Begründer der messenden Experimentalphysiologie 1816–1895*. Stuttgart, 1967.

Schröter, Harm G. "Kartelierung und Dekartalierung 1890–1990." *Vierteljahrsschrift für Sozial- und Wirtschaftsgeschichte* 81 (1994): 457–93.

Scott, Joan W. "The Class We Lost?" *International Labor and Working Class History* 5 (2000): 69–75.

Šedivý, Ivan. *Češi, české země a velká válka 1914–1918*. Prague, 2001.

Shorter, Edward. *Women's Bodies: A Social History of Women´s Encounter With Health, Ill-Health, and Medicine*. New Brunswick, 1991.

Showalter, Elaine. "Rivers and Sassoon: The Inscription of Male Gender Anxieties." In *Behind the Lines: Gender and the two World Wars, edited by* Margaret R. Higonnet, Jane Jenson, Sonya Michel, and Margaret Collins-Weitz, New Heaven, CT and London, 1987, 61–69.

Sieder, Reinhard J. "Behind the Lines: Working-Class Family Life in Wartime Vienna." In *The Upheaval of War: Family, Work and Welfare in Europe, 1914–1918*, edited by Richard Wall and Jay Winter, Cambridge, 1988, 109–138.

Šimůnková, Alena. "Statut, odpovědnost a láska: vztahy mezi mužem a ženou v české měšťanské společnosti v 19. století." *Český časopis historický* 95 (1997): 55–107.

Smrček, Otto. "Expanze technického myšlení." In *Technokracie v českých zemích (1900–1950)*, edited by Jan Janko and Emilie Těšínská, Prague, 1999, 37–56.

Šolle, Zdeněk. *Dělnické stávky v Čechách v druhé polovině 19. století*. Prague, 1960.

———. "K počátkům dělnického hnutí v Praze." *Československý časopis historický* 5 (1957): 664–87.

Srb, František et al. *Nástin dějin Československého odborového hnutí. Od vzniku prvních organizací odborového typu do období nástupu k výstavbě socialismu*. Prague, 1963.

Stedman-Jones, Garreth. *Outcast London: A Study in the Relationship Between Classes in Victorian Society*. Oxford, 1971.

Stollberg, Gunnar. *Die Rationalisierungsdebatte 1908–1933: Freie Gewerkschaften zwischen Mitwirkung und Gegenwehr*. Frankfurt am Main and New York, 1981.

Sun, Raymond C. "Hammer Blows: Work, the Workplace, and the Culture of Maculinity Among Catholic Workers in the Weimar Republic." *Central European History* 2 (2004): 245–71.

Svobodová, Jiřina. "Rodina a rodinný život pražského dělnictva. " In *Stará dělnická Prague. Život a kultura pražských dělníků 1848–1939*, edited by Antonín Robek, Mirjam Moravcová, and Jarmila Šťastná, Prague, 1981, 107–36.

Sylbey, David. "Bodies and Cultures Collide: Enlistment, the Medical Exam, and the British Working Class, 1914–1916." *Social History of Medicine* 17, no. 1 (2004): 61–76.

Tanner, Jakob. *Fabrikmahlzeit, Ernährungswissenschaft, Industriearbeit und Volksernährung in der Schweiz 1890–1950*. Zürich, 1999.

Těšínská, Emilie, and Jindřich Schwippel. "Masarykova akademie práce." In *Bohemia Docta. K historickým kořenům vědy v českých zemích*, edited by Alena Míšková, Martin Franc, and Antonín Kostlán, Prague, 2010, 286–331.

Thomas, Sven. *Gustav Schmoller und die deutsche Sozialpolitik*. Düsseldorf, 1995.

Thompson, Edward P. *The Making of the English Working Class*. London, 1961.

———. "The Moral Economy of the English Crowd in the Eighteenth Century." *Past and Present* 50, no. 1 (1971): 76–136.

Thoms, Ulrike. "Essen in der Arbeitswelt. Das betriebliche Kantinenwesen seit seiner Entstehung." In *Die Revolution am Esstisch: Neue Studien zur Nahrungskultur im 19./20. Jahrhundert*, edited by Hans J. Teuteberg, Stuttgart, 2004, 203–218.

Tilly, Chris, and Charles Tilly. *Work under Capitalism*. Oxford, 1997.

Tinková, Daniela. *Tělo, věda, stát: zrození porodnice v osvícenské Evropě*. Prague, 2010.

Todd, Lisa M. "'The Soldier's Wife Who Ran Away with the Russian': Sexual Infidelities in Wolrd War I Germany." *Central European History* 44 (2011): 257–78.

Troch, Harald. *Rebellensonntag. Der 1. Mai zwischen Politik, Arbeiterkultur und Volksfest in Österreich (1890–1918)*. Vienna, 1991.

Unfried, Berthold. "Arbeiterprotest und Arbeiterbewegung in Österreich während des Ersten Weltkrieges." Ph.D. diss., University of Vienna, 1990.

Unowsky, Daniel L. *The Pomp and Politics of Patriotism: Imperial Celebrations in Habsburg Austria 1848–1916*. West Laffayette, 2005.

Urban, Otto. *Česká společnost 1848–1918*. Prague, 1982.

Urbanitsch, Peter. "Mýtus pluralismu a realita nacionalismu. Dynastický mýtus habsburské monarchie—zbytečná snaha o vytvoření identity?" *Kuděj* 1 (2006): 46–68; and 2 (2006): 35–53.

Urzidil, Johannes. *Pražský triptych.* Prague, 1997.

van der Linden, Marcel, and Jürgen Rojahn, eds. *The Formation of Labour Movements 1870–1914: An International Perspective,* vols. I–II. Leiden, 1990.

van Laak, Dirk. "Planung. Geschichte und Gegenwart des Vorgriffs auf die Zukunft." *Geschichte und Gesellschaft* 34, no. 3 (2008): 305–26.

Vatin, François. "Arbeit und Ermüdung. Entstehung und Scheitern der Psychophysiologie der Arbeit." In *Physiologie und industrielle Gesellschaft. Studien zur Verwissenschaftlichung des Körpers im 19. und 20. Jahrhundert,* edited by Philipp Sarasin and Jakob Tanner, Frankfurt am Main, 1998, 347–368.

Vernon, James. *Politics and the People: A Study in English Political Culture c. 1815–1867.* Cambridge, 1993.

Voss, Lex Heerma van, and Marcel van der Linden, eds. *Class and Other Identities: Gender, Religion and Ethnicity in the Writing of European Labor History.* New York, 2002.

Vrints, Antoon. "Beyond Victimization: Contentious Food Politics in Belgium During World War I." *European History Quarterly* 45, no. 1 (2015): 83–107.

Wagner, Peter. *A Sociology of Modernity: Liberty and Discipline.* London and New York, 1994.

Weber, Max. *Die protestantische Etik und der Geist des Kapitalismus.* Tübingen, 1934.

Wedel, Michael, ed. *Max Mack: Showman im Glashaus.* Berlin, 1996.

Welskopp, Thomas. "Atbeitergeschichte im Jahr 2000. Bilanz und Perspektive." *Traverse* 2 (2000): 15–31.

———. "Von der Verhinderten Heldengeschichte des Proletariats zur Vergleichenden Sozialgeschichte der Arbeiterschaft—Perspektiven der Arbeitergeschichtsschreibung in den 1990er Jahren." *Zeitschrift für Sozialgeschichte des 20. und 21. Jahrhunderts* 3 (1993): 34–53.

Whitley, Richard. "Knowledge Producers and Knowledge Acquirers: Popularization as a Relation Between Scientific Fields And Their Publics." In *Expository Science: Forms and Functions of Popularisation. Sociology of the Sciences,* edited by Terry Shinn and Richard Whitley, vol. IX., 1985, 3–28.

Wingfield, Nancy M. "The Enemy Withinn: Regulating Prostitution and Controlling Venereal Disease in Cisleithanian Austria during the Great War." *Central European History* 46 (2013): 568–98.

Wingfield, Nancy M., and Bucur, Maria. "Introduction: Gender and War in Twentieth Century Eastern Europe." In *Gender and War in Twentieth Century Eastern Europe,* edited by Nancy M. Wingfield and Maria Bucur, Bloomington, IN, 2006, 1–20.

Winkler, Wilhelm. *Die Einkommensverschiebungen in Österreich während des Weltkrieges.* Vienna and New Haven, CT, 1930.

Zahra, Tara. *Kidnapped Souls: National Indifference and the Battle for Childern in the Bohenian Lands 1900–1948.* Ithaca, 2008.

Zückert, Martin. "Der erste Weltkrieg in der tschechischen Geschichtsschreibung 1918–1938." In *Geschichtsschreibung zu den böhmischen Ländern im 20. Jahrhundert,* edited by Christiane Brenner, K. Erik Franzen, Peter Haslinger, and Robert Luft, München, 2006, 61–75.

Zunkel, Friedrich. *Industrie und Staatssozialismus. Der Kampf um die Wirtschaftsordnung in Deutschland 1914–1918*. Düsseldorf, 1974.

Musical and Film Recordings

Bad Religion. *Generator.* Epitaph Records, 1992.

Das Stahlwerk der Poldihütte während des Weltkrieges, Sascha-Messter-Film, 1916.

Index

Abderhalden, Emil, 20–21. *See also* physiology
academic knowledge, 2, 63
academics, 43, 67
Adámek, Karel, 68
affluence, 28
army, 68, 70, 72–74, 80, 102, 105–6, 109–10, 113, 116–17, 132, 144
Austria, 43, 46, 63, 68, 73, 82, 101–2, 108, 131, 139, 142–43, 156
  Austria-Hungary, 2, 6, 8, 18–19, 27
  Austrian Alpine Company, 81
  Austrian business inspection, 44
  Austrian cinematography, 61. *See also Das Stahlwerk der Poldihütte* (The Poldi Steelworks) (film)
  Austrian government, 19, 131, 133
  Austrian heavy industry. *See* industry
  Austrian justice system, 2, 135
  Austro-Hungarian Empire, 6, 67, 79
  Austro-Hungarian political power, 7

Baden, 137
Balkan Wars, 69
Bavarian workers. *See* workers
Benedict, Francis Gano, 26. *See also* physiology
Beneke, Friedrich Wilhelm, 21. *See also* physiology
Beneš, Edvard, 28
Benešov, 113
Benešová, Hana, 28
Berger, Vojtěch, 30, 105, 116
Berlin, 21, 38, 67, 94
Bílina, 46
biology, 21, 31
  biological sex, 115
biomechanics, 16
bodily functions, 16, 21
Bohemia, 24, 34, 43–44, 47, 72, 84, 115, 142, 165
  Bohemian lands, 3–5, 8, 14–15, 19–20, 22, 27–28, 43–44, 46–47, 68, 70, 73, 86, 95, 97–98, 101, 133–35, 143, 146, 152, 154
  Central Bohemia, 44
Bohemian Institute of Agronomy, 21. *See also* Stoklasa, Julius
Bolevec, 144
„Borská pole", 145–48. *See also* Hofmann, Josef
Bory, 140
Boskovice, 83
Brno, 28–29, 108, 113, 120
Broadway, 94
Brož, Václav, 40
Brussels, 67
Bubny, 152

Bystrc, 83

Café Waldek, 131
Caha, Ladislav, 103
calories, 1, 21–23, 38–39, 45, 111
caloric values, 21–22
Canetti, Elias, 149
capital, 27, 30, 64, 78–80, 141
carbohydrates, 21–22, 25, 32–33, 39
censorship, 81–82
České Budějovice (Budweis), 47, 73–74
Český Brod, 44
Charles University, 20. *See also* Mareš, František
chemical analyses, 26
Chittenden, Russel H., 17, 21, 24. *See also* physiology
Chlupatá, Žofie, 13–14
Chlupatý, František, 13–14, 36
Chomutov, 46
Christian morality, 15, 64
Christian theology, 64
Christmas, 29, 36–37
city, 4, 23, 28–29, 34–35, 38–39, 41, 44–45, 98–99, 107, 110, 130–31, 138–39, 141–42, 152–53, 155–56, 164, 168
Clausius, Rudolf, 15–16. *See also* thermodynamics
collective action, 6, 8, 79, 102, 118, 134, 165
collective identity, 3, 6, 48, 85
collective negotiations, 70, 76, 85
Communist Party, 3, 86
Czechoslovak Communist party, 169
communist utopia, 3, 86
complaint committees, 70
Connell, Raewyn W., 116
consumption, 2, 5, 7–8, 13, 15, 19, 24–26, 29, 33, 35, 42, 45, 48, 96, 98–99, 111, 157, 164
food consumption. (*see* food)
meat consumption, 13–14, 23–24, 34
Prague committee for public consumption. (*see* Prague)
cooking, 13, 31–32, 38, 40, 97–98, 110
countryside, 6, 28–29, 37, 141–42, 167
Ctiborová, Marie, 105
culture, 6, 31, 104, 118, 140–41
cultural history, 4
popular culture, 25
working-class culture. (*see* working-class)
working-class domestic culture. (*see* working-class)
Czech Association of Unions, 34
Czech federation of metalworkers' unions, 34
Czech history, 4, 101
Czech National Socialist Party, 169. *See also* Klofáč, Václav
Czech newspapers, 23, 25
Czech physiology. *See* physiology
Czech Social Democratic party. *See* Social Democratic party
Czech Technical University, 69
Czech workers. *See* workers
Czech working class. *See* working class
Czechoslovak Communist Party. *See* Communist party
Czechoslovak Republic, 86
Czechoslovakia, 3, 169
Czechs, 6, 23

Daimler Motoren Company in Wiener Neustadt, 76
Das Stahlwerk der Poldihütte (The Poldi Steelworks) (film), 61–62, 76. *See also* Messter, Oskar and „Österreichisch-Ungarische Sascha-Messter-Filmfabrik GmbH"
Dělnické listy, 29
Die Welt ohne Männer (film), 94–95
diet, 17, 23–24, 26, 28, 45, 77, 82, 97, 112
diet specialists, 16, 18, 20, 23, 26, 39
dietary science. (*see* science)
working class diet. (*see* working-class)

disciplinary regime, 14, 115–116
disciplining of subjects, 3
discourse, 3, 6, 15, 17, 26–27, 31, 33–35, 37–38, 45, 48–49, 62, 64–65, 68–69, 75, 77–78, 80–82, 84–85, 96–98, 103–5, 110–13, 117, 138, 164–65
Domes, Franz, 145–46. *See also* parliament
Donát, Josef, 153
Doudlevce, 140

eating habits, 21, 23
economy, 24, 33, 47, 80–81, 96, 100, 113, 170
  economic growth, 96
  economic production, 5, 48, 72
  economic productivity, 63, 68
  liberal economic order, 15
  „moral economy", 137–38, 142, 167. (*see also* Thompson, Edward Palmer)
  „rucksack economy" (Rucksackwirtschaft), 28
  wartime economy. (*see* wartime)
effectiveness, 17–18, 62, 64, 68–69, 71, 73–74, 76, 78
efficiency, 17, 97
emotionality, 134–35
emperor, 106
employees, 27, 37, 44, 47, 69, 72, 74, 76, 78, 108, 113–14, 119–20, 132, 146, 153
employers, 30, 34, 48, 69–71, 74, 76–78, 80, 82, 84–85, 87, 111–12, 118, 134, 150, 164–66, 169
energy, 1, 15–18, 21–23, 25–26, 31–33, 38, 40, 45–48, 62, 67–69, 77, 84, 96, 104, 111, 114
  energy intake and release, 16
  energy transfer, 15, 16, 31, 45, 164 (*see also* Clausius, Rudolf and Helmholtz, Hermann von)
  kinetic energy, 17
  thermal energy, 16, 21
Engel, Alexander, 94. *See also Die Welt ohne Männer* (film)
Engel, Ernst, 111
England, 22
Entropy, 15
ergograph, 65–66. *See also* Mosso, Angelo
Erlangen, 66. *See also* Weichardt, Wilhelm
eugenic medicine, 103
Europe, 15–18, 21–22, 67–68, 102, 111
  Central Europe, 4
  East-Central Europe, 4
  European Enlightenment, 2
  European historiographies. (*see* historiography)
  European industrialization, 5, 97, 112, 138 (*see also* industry)
  European physiology (*see* physiology)
everyday life, 14, 18, 20, 29, 95, 97–99, 110
exertion, 22, 64, 66, 69
exhaustion, 66, 114
  bodily exhaustion, 1, 65
  total exhaustion, 1, 17

factory, 3, 12, 14, 17, 27, 33–34, 38, 42–44, 46–49, 57–59, 61–64, 69–78, 80–81, 84, 86, 110–15, 117, 120, 131, 134, 137, 139–47, 150–56, 165, 167
  factory hierarchy, 76
family, 7, 12, 14, 25, 28–31, 33–37, 44, 63, 72, 80, 98–99, 101, 106, 110, 118, 121, 140, 146, 148, 166
farmer, 29–30, 37, 141
fat, 21–23, 25, 32–33, 39, 46, 49
fatigue, 17, 56, 62–69, 75, 77, 121
Fatigue (La Fatica) (book), 66. *See also* Mosso, Angelo
femininity, 96
Fleischner, Jindřich, 69
Florisdorf, 81
food, 1, 7–8, 14–15, 17–40, 44–48, 62–65, 68–69, 77, 82–83, 96–99, 111, 121, 136–39, 142, 144–49, 164–65, 170
  consumption of food, 7, 14, 21–22, 27, 35–36, 48, 62, 95–96, 98–99, 112

distribution of food, 7, 19, 35, 45, 131, 133, 149
food line, 36, 44, 49, 130, 138, 141, 164
food rations, 1, 20–21, 25–27, 36–37, 44, 81, 96, 130, 166
food shortage, 14, 23, 26, 34, 45, 134, 139, 164
politics of food, 7, 12, 15, 31, 35, 48, 62, 77, 96, 163, 165
Forster, Georg, 21. *See also* physiology
France, 22
František Křižík electrotechnical factories, 152–53, 155
Franz Joseph I, 106
front, 35, 72, 75, 77, 83–84, 96, 100–105, 108–109, 117, 130, 133, 141, 145, 147, 149, 163
home front, 48, 100–105, 109–110, 115–17, 121

Galandauer, Jan, 3
Gautier, Armand, 22. *See also* physiology
gender, 4, 7–8, 66, 96–97, 100–102, 106, 110, 112, 114–16, 122, 134, 141, 149, 154–55, 157, 166–69
gender difference, 97
gender diversification, 8, 110, 121
gender order, 95, 99–100, 108, 110, 115–16, 121, 166
gender relations, 95–96, 99, 116, 121, 157
gender relationships, (*see* gender relations)
politics of gender, 94, 96
Germany, 18, 63, 67, 101
German cinematography, 61
see also Stahlfabrik Krupp (The Krupp Steelworks) (film) and *Thomaswerk* (The Thomas Works) (film)
German-speaking citizens, 6
gonorrhea, 115
government, 14, 18–19, 27, 29–30, 46, 63, 70, 78, 99, 105, 112, 131, 133, 136, 138, 145, 147, 153
Gurný, Josef, 136

Habrman, Gustav, 104, 148. *See also* Czech Social Democratic party
Habsburg Monarchy, 2, 4, 7, 14, 18–19, 21, 24, 48, 61, 69, 72, 100, 106, 114, 130, 136, 141, 163, 167, 169
Hampl, Antonín, 84
harvest, 27–28, 46
Hasselsteiner, Horst, 4
Healy, Maureen, 4, 19, 102
Helmholtz, Hermann von, 15–16. *See also* thermodynamics
Heumos, Peter, 133
Hindhede, Mikkel, 17, 21, 24. *See also* physiology
hinterland, 2, 100, 102, 131, 154
history of social protest, 4
historiography, 5, 86, 100–102, 133–35, 141–43
American historiographies, 4
Czech historiography, 101, 134, 143
European historiographies, 4
social historiography, 31
socialist historiography, 3
Western historiography, 4
Hložek, František, 7. *See also* Poldi factory
Hofmann, Josef, 148
Holešovice, 38–39, 41–42, 152
Horst, Julius, 94. *See also Die Welt ohne Männer* (film)
Hradec Králové (Königgrätz), 93
human, 2, 5, 7, 15–18, 20–21, 24–25, 28, 30, 33, 45, 47, 57–65, 67–69, 75, 77, 99, 111
human body, 16–17, 21–22, 26, 28, 31–33, 39, 48, 62, 64, 66–69, 77, 96
human motor, 16, 18, 21, 26, 30, 38–39, 45–46, 62, 64, 69–70, 74, 165. (*see also* Amar, Jules)

human resources management. *See* management
Hungary, 34
hunger, 24, 27, 33, 82, 133–34, 136–37, 148
hygiene, 21, 33, 111, 140

income, 19, 28, 30, 33, 35–36, 43, 114, 151, 163
industry, 34, 111, 116, 131, 153
  Austrian heavy industry, 62, 76
  heavy industry, 34, 119
  machine-building industry, 72
  militarized industry, 1
  textile industry, 34
  wartime industry. (*see* wartime)
inflation, 19, 144
injury, 61
insurance, 82
  accident insurance, 68
  health insurance, 114
  medical insurance, 68
Ioteyko, Josefa, 67
iron, 57–59, 61, 82, 102, 113, 117

Jelínek, Ludvík, 12–14
Jičín, 83
Julianov, 83

Kacafírek, union treasurer, 83
Kafka, Josef, 24–25
Kantorowicz, Ernst, 106
Karabatschek, Hans von, 76, 144
Karlín, 44, 152–53. *See also* František Křižík electrotechnical factories
Karlovy Vary, 46
Kárník, Zdeněk, 3
Kejřová, Anuše, 32, 97–98
Kellner, Ferdinand, 39
kitchen, 12, 25, 36, 38–45, 137, 164
  community kitchen, 34
  „middle-class" kitchens (*see* middle-class)
  people's kitchens, 41–43
  public kitchens, 13, 35, 37–45, 164
Kladno, 71, 136, 142
Klofáč, Václav, 155. *See also* Czech National Socialist Party
Kolben electrotechnical company, 152
Kovodělník, 77, 81–82, 116
Kraepelin, Emil, 68
Kratochvíl, Adolf, 153, 155
Krejčí, Josef, 153, 155–56
Kühn, military administrator of the Pilsen Škoda factory, 71, 144, 146

labor, 1, 3–4, 6–8, 14–18, 22–25, 27, 30, 34, 45, 48, 60, 62–64, 67–71, 75–81, 83–85, 96, 99, 102, 111–12, 114–18, 121–22, 140, 164–67
  child labor, 5
  „labor aristocracy", 79
  labor code, 68
  labor history, 3–4, 102
  labor market, 5, 165
  labor unions, 34, 79, 116–21
  physical labor, 7, 26
labor code. *See* labor
labor unions. *See* labor
Lagrange, Ferdinand, 67
language, 4, 15–16, 20, 27, 36, 63, 67, 79, 118, 141, 145–47, 151–52, 154–55
lathe, 30, 59, 62, 113, 153, 155
Laurin and Klement car company, 137
law, 3, 13, 16, 18, 43, 64, 66, 69–70, 73, 77–79, 102, 112, 114–15
  criminal law, 70
  law on state protected enterprises, 153
  „Law on Wartime Operations", 69, 78, 102
  laws of thermodynamics. (*see* thermodynamics)
  lawyers, 42–43, 138
  martial law, 70, 130, 132–33
  public law, 70
  trade law, 74
„Law on Wartime Operations". *See* law
Lessing, Madge, 94. *See also Die Welt ohne Männer* (film)
Libeň, 12–13, 41–42, 137, 152

liberal economic order. *See* economy
liberal order, 85, 87, 138–40
Liberec (Reichenberg), 137
Litoměřice, 72, 113
living conditions, 63
Lobzy, 140
local administration, 34, 44, 147
Lochotín, 146–47
Loket, 46
London, 94
looting, 131–32, 135–39, 142, 168
Ludwig, Carl, 16. *See also* physiology
luxury, 14, 28, 37, 131

Mack, Max, 94. *See also Die Welt ohne Männer* (film)
machine, 15–17, 21, 25, 32, 33–34, 47, 57, 59, 61–62, 64, 71, 73, 76–77, 81, 96, 113, 117, 146, 150, 153
male, 6, 22, 42, 72, 95–101, 103, 106–118, 120–22, 134–36, 148, 154–57, 166–67, 169
  male authority, 7, 101, 110, 122, 166
  male breadwinner, 99
  male hegemony, 7
management, 38, 47, 64, 73, 75–76, 83, 97–98, 140, 144–45, 150
  human resources management, 75
manhood, 103, 105, 121
manliness, 94
Mareš, František, 20, 24, 26. *See also* physiology
market, 15, 19, 25, 29, 32, 34, 38, 64, 85, 137–38, 167
  black market, 30
  free market, 5
  labor market. (*see* labor)
Marschner, Robert, 100
martial law. *See* law
Marxist, 4, 86, 169
masculinity, 96, 101–106, 110, 112, 114, 116, 121, 167
  collapse of masculinity, 101
  masculine authority, 108
material shortage, 6
Matula, Robert, 99
meatless days, 13
mechanical power, 18
Merhaut, Antonín, 20, 23
Messter, Oskar, 61, 76. *Das Stahlwerk der Poldihütte* (The Poldi Steelworks) (film)
metalworkers. *See* workers
middle class, 24–25, 27, 31, 36, 41–42, 49, 63, 76, 97–98, 136, 138–42, 168
  middle-class kitchens, 40–43, 49, 164
  middle-class professions, 43
military, 44, 46–47, 70, 72–73, 78, 82, 85, 100, 102–103, 105–106, 109–110, 113, 117, 132, 137, 140, 142, 144–47, 151, 153–55, 169
  military administrator, 45–46, 71–73, 75, 112–15, 120, 137, 144, 146
  military draft, 77, 84
  military service, 102–104
miners′ unions, 79, 83
  Union of Austrian Miners, 79
Ministry of War, 45
misery, 28, 100
Mladá Boleslav (Jungbunzlau), 137. *See also* Laurin and Klement car company
mobilization, 2, 26, 69–70, 83, 107, 153
modernity, 2, 61
monarchic power, 106
Moravia, 34
mortality, 63
Mosso, Angelo, 65–67. *See also* ergograph
Most (Brüx), 46
muscle machine, 21
Museum of the Kingdom of Bohemia, 24
Mutějovice, 28

Národní Listy, 20, 23, 75
naval blockade, 27
Nedvěd, Antonín, 28
Neruda, Jan, 150

non-German ethnic groups, 6
Nová Doba, 30, 33, 78, 107
nutrition, 15–18, 20–21, 23–24, 26, 30, 32–34, 39, 46–47, 97, 111, 165
    malnutrition, 27–28, 30, 115
    nutritional guidelines, 45
    nutritional science, 22–25, 27, 98, 164
    rational nutrition, 23, 25–26, 32

organization
    party organization, 4
organized capitalism, 80, 87
Ostrava, 99
output, 16–18, 68, 73, 75, 165
    work output. (*see* work)
„Österreichisch-Ungarische Sascha-Messter-Filmfabrik GmbH", 61. *See also Das Stahlwerk der Poldihütte* (The Poldi Steelworks) (film)

Pacovský, Oldřich, 78
Pardubice, 44, 73–74, 113
Paris, 67, 94
parliament, 79, 121, 145–46, 169. *See also* Domes, Franz and Pik, Luděk
    parliamentary politics. (*see* politics)
Pazourek, Josef, 69
peasants, 30
Peklo, workers' house in Pilsen, 94–95, 145
Pettenkoffer, Max Josef von, 17, 23
physical strength, 1, 47, 84, 135
physics, 15, 26
physiology, 16, 20, 22, 24, 27, 31, 65–67
    Czech physiology, 68
    European physiology, 66
    physiological and dietary knowledge, 25 (*see also* diet)
    physiological experiments, 26
    physiologist, 16, 18, 20–22, 24–26, 39, 65, 67
Pik, Luděk, 145–46. *See also* parliament
Pilsen, 28–30, 33–34, 37, 44, 71–73, 76, 78, 81–82, 94–95, 107–108, 110, 113–14, 130–33, 135–36, 138–42, 144–48, 150–57, 164, 168
planning, 2, 63, 67
    social planning, 2
Plaschka, Georg, 4
Plavec, Josef, 1–2
Playfair, Lyon, 22
Poldi factory, 71. *See also* Hložek, František
police raid, 13
politics, 4, 6–8, 15, 63, 87, 96, 142, 165, 167, 169
    parliamentary politics, 6
    political system, 6
    political party, 3, 79, 116
    „politics of food". (*see* food)
    politics of gender. (*see* gender)
Popelka, J., 39
popular culture. *See* culture
poverty, 14, 19, 38, 63, 141, 148, 163
Prager Tagblatt, 75
Prague, 1, 12–14, 20–21, 23, 29, 31, 34–35, 37–44, 67, 70, 73, 81, 84, 98, 107, 113, 132, 137, 140, 144, 147, 150, 152–57, 164, 168
    Prague committee for public consumption, 39
    Prague council, 40
    Prague municipal office, 13–14
Právo Lidu, 34, 36–37
price, 13–14, 19, 29–32, 34, 38, 40–42, 44–45, 77, 138
    „fair prices", 138, 142, 167
private ownership, 85–86, 168
proletarianization, 63, 140
propaganda, 20, 102
protein, 21–26, 32–33, 39, 46
protest, 4, 7–8, 44, 130–31, 133–38, 141–43, 146–48, 150–52, 154–56, 164, 167–68
    workers' protest. (*see* workers)
public interest, 19, 99
public space, 7, 27, 29, 47, 72, 81, 98, 100–101, 103–104, 106, 108–110, 117, 166

publicly organized motherhood, 96–97, 99
Pulec, union treasurer in Boskovice, 83

Rabinbach, Anson, 17
rank and file, 3, 45, 169
rationality, 64, 69, 95, 134–35, 139, 166, 168
ration, 2, 19–20, 23, 36–37, 47, 96, 121, 130, 137, 141, 163, 166, 170
 food rations. (*see* food)
 ration card, 14, 36–37, 44, 47, 137
 rationalization, 2, 18–19, 27, 31, 38, 61, 64, 69, 75–78, 84–86, 96–98, 112, 150, 165, 170
 rationing, 2, 7, 165
resistance, 12, 70, 118, 120–21
Rettigová, Magdalena Dobromila, 98
revolution, 63, 86, 99, 143
 Russian Revolution, 86
rights, 3, 30, 78, 95, 121–22, 151, 167
 civil rights, 2, 6, 70
 liberal rights, 6
Richet, Charles, 67
Ringhoffer factory, 1, 70, 152–53, 155–56
Ringhoffer railway coach company. *See* Ringhoffer factory
Ringhoffer train car company. *See* Ringhoffer factory
riot, 132–33, 135, 137, 139, 142, 167–68, 170
Rubner, Max, 21, 39

sabotage, 1
salary, 5, 15, 19, 28–29, 30, 33–34, 45, 48, 64, 73, 76, 79–82, 147–48, 151, 163
Sarasin, Philipp, 18
science, 2–3, 7, 16, 18, 20, 26–27, 62, 64–65, 76, 85, 96
 dietary science, 26–27, 31–32
 modern science, 7, 18, 26–27, 48–49, 63, 65, 97
 natural science, 15–16, 18, 20, 164
 scientific knowledge, 2–3, 7, 17, 20, 27, 31, 64, 75, 99, 163, 165–66
 technical sciences, 76
Šedivý, Ivan, 4, 101, 134
sexual morals, 115
Singer, Richard, 13
Skaunicová, Františka, 120
Skvrňany, 140
Škoda, Emil, 117
Škoda factory, 44, 71–73, 76, 78, 81–82, 113–14, 132, 140–41, 144–48, 150–51, 153, 155, 168
Slaný, 136, 142
slavery, 5
Smíchov, 113, 152–53, 156
Smolík, Rudolf, 39
Social Democratic party, 3–4, 34, 78–79, 152, 169
 Czech Social Democratic party, 3, 104. (*see also* Habrman, Gustav)
socialist dictatorships, 4
socialist historiography. *See* historiography
society, 2–4, 6, 20, 26–28, 35, 40, 42–43, 48, 95, 100, 102, 106, 138–39, 163, 168, 170
 industrializing societies, 63
 social conditions, 63
 social conflicts, 2, 63
 social hierarchy, 6, 28, 37, 48, 164
 social historiography. (*see* historiography)
 social problem, 35, 63, 140
 social tensions, 6
 „social question", 63–64, 97
 Western society, 3
Sokolov, 46
standard of living, 5, 33, 35, 38, 140
Stahlfabrik Krupp (The Krupp Steelworks) (film), 62
state, 1, 3, 6–8, 13–17, 19, 29–30, 33–34, 43–45, 47–48, 68, 70–71, 73–75, 77, 79, 81–82, 84–86, 106, 108, 119–21, 130–31, 133–34, 147, 149, 152–54, 163–64, 166–67, 170
statistics, 63, 82, 102, 114
steel, 58–61, 136
 steelworks, 58, 61

Stoklasa, Julius, 21, 23–24, 26, 39, 136. *See also* physiology
strike, 1–2, 27, 44, 78, 80, 131, 133–34, 136–57, 167–69
suffering, 28, 36–37, 63, 82, 100, 115, 121, 138, 148, 151
Suk, R. A., 98
Šulc, Jaroslav, 153, 155
Suppan, Arnold, 4
supply crisis, 19, 28, 33, 37, 39–40, 47, 96, 136, 163
symbols, 5–6, 63, 107, 110, 122, 138–40, 143, 150
syphilis, 115
Syrovátka, manager of the Prague public kitchen for the poor, 40

Tandler, Julius, 103
Tanner, Jakob, 17, 111
Taylor, Frederik Winslow, 64
technology, 20, 117
Teplice, 46–47, 135
thermodynamics, 18
    laws of thermodynamics, 15, 17. (*see also* Clausius, Rudolf and Helmholtz, Hermann von)
Thomaswerk (The Thomas Works) (film), 62
Thompson, Edward Palmer, 137. *See also* economy
trade inspection, 73, 113
trade law. *See* law
Trutnov, 46, 113
tsarism, 104
    Russian tsarism, 104

U Palmů restaurant, 41
U Vejvodů restaurant, 41–42
unemployment, 33–35, 38, 79, 84
Union of Metal Workers, 78, 80–81
Union of Railway Employees, 83
Union of Railway Workers (*see* Union of Railway Employees)
United Nations, 22
University of Halle, 20. *See also* Abderhalden, Emil.
University of Munich, 22. *See also* Voit, Carl von
unrest, 27, 47, 130–33, 135–37, 139, 142–44, 146, 152–54, 156, 167–68
urban population, 28, 37
Ústí nad Labem, 46

venereal diseases, 115
Vienna, 4, 19, 38, 45–46, 61, 67, 76, 102, 113, 132, 137, 144, 147, 153
Vinohrady, 113
violence, 4, 71, 130–32, 134, 139, 142, 150, 155–56, 165, 170
Vítkovice, 136
    Vítkovice steelworks, 76
Vlašim, 113
Voit, Carl von, 22–23, 25, 111. *See also* physiology
Vysočany, 152

war returnees, 80
wartime, 2–8, 15, 19–24, 26–29, 31, 35, 43, 46, 48–49, 61–62, 69–70, 73–84, 86–87, 95–96, 99–103, 106, 110–12, 115, 118, 130–31, 133, 137, 139, 140–44, 154, 156–57, 163–65, 167–69
    wartime economy, 2, 6, 23, 26, 28–30, 43, 69, 77, 80–83, 86, 165
    wartime industry, 7, 71, 81, 85, 102, 110–12, 114, 118, 131, 152, 165
    wartime industrial production, 15
    wartime scarcity, 7, 23
Washington, 67
Weber, Max, 2
Weberian approach, 3
Weichardt, Wilhelm, 66, 68. *See also* physiology
Western historiography. *See* historiography
Western society. *See* society
Weyr, František, 28
work, 1–5, 13, 15–23, 25–26, 31–35, 38–39, 45–48, 57, 59, 61–62, 64, 66–71, 73–78, 81, 84–87, 96–97, 102–104,

109–117, 121, 135, 137, 139, 141–55, 157, 163, 165–67, 169
  gainful work, 5
  industrial work, 4, 17, 96, 114, 117–18
  productive work, 5, 111
  reproductive work, 36, 97, 110–111
  work breaks, 68, 73–74
  work output, 17–18, 22–23, 26, 32–33, 62
  work relationships, 75, 77
  working conditions, 33, 70, 110, 144
  working day, 17, 33, 74, 84
worker, 1–8, 13–14, 22–24, 27, 29–31, 33–35, 37–40, 42–49, 58–64, 67–82, 84–86, 94–97, 99, 103–106, 109–110, 112–22, 132, 134–57, 163–69
  Bavarian workers, 22
  Czech workers, 2, 5–6, 86, 104, 141
  metalworkers. (*see also* Union of Metal Workers)
  workers' collective, 7, 80, 118, 141
  workers' newspapers, 33, 35, 79, 81–82, 95
  workers' protest, 7–8, 44, 142, 148, 152, 154, 157
working class, 8, 25, 29–37, 41–42, 63, 77–80, 82–85, 95–99, 102, 104–105, 109, 110–12, 114, 116–21, 139–40, 142, 147, 150–52, 156–57, 163–69
  Czech working class, 3, 5, 86, 99, 122
  working-class culture, 30–31, 33, 75, 77, 79, 84–85, 99, 116, 118, 121, 147–48, 164–66, 169
  working-class diet, 45, 62
  working-class domestic culture (*see* working-class culture)
  working-class identity, 79
World War I, 3–5, 14, 16–19, 31, 48, 62–63, 65, 68–69, 76–77, 94–97, 100–102, 104, 106, 110–11, 116, 121–22, 134, 137, 141, 143, 168–70
woman, 6–7, 14, 22, 25, 28, 36–37, 48, 59, 76, 94–95, 97–100, 105, 107–122, 130–32, 134–37, 139, 141, 146–48, 150–51, 155–56, 166–67, 169
  women's education, 97
  women's emancipation, 95, 120
  women's emancipation movement, 95

Zájmy žen, 98, 121
Žižka, František, 13–14, 36
Žižkov, 37, 113
Žižková, Anna, 13–14
Zuntz, Nathan, 67. *See also* physiology